International Science

Coursebook 1

international Science

Coursebook 1

Karen Morrison

HODDER EDUCATION
AN HACHETTE UK COMPANY

The Publishers would like to thank the following for permission to reproduce copyright material:

Photo credits
All photos supplied by Mike Van Der Wolk except for **p.18** Fig 2.2b Graham Jordan/Science Photo Library;
p.20 Fig 2.5c © Gallo Images/Corbis; **p.25** Fig 2.12 Holt Studios International/Alamy; **p.36** Fig 3.10 Biophoto
Associates/Science Photo Library; **p.58** Fig 5.4. © Micheal Sewell/Still Pictures; **p.59** Fig 5.6 Rod Planck/Science
Photo Library; **p.63** Fig 5.13a Jacques Jangoux/Alamy; **p.159** Courtesy of Thebe Medupe; **p.172** Fig 14.16
Frederic J. Brown/AFP/Getty Images

Every effort has been made to trace all copyright holders. If any have been inadvertently overlooked,
the Publishers will be pleased to make the necessary arrangements at the first opportunity.

Hachette UK's policy is to use papers that are natural, renewable and recyclable products and made from wood
grown in sustainable forests. The logging and manufacturing processes are expected to conform to the environmental
regulations of the country of origin.

Orders: please contact Bookpoint Ltd, 130 Milton Park, Abingdon, Oxon OX14 4SB. Telephone: (44) 01235 827720.
Fax: (44) 01235 400454. Lines are open 9.00–5.00, Monday to Saturday, with a 24-hour message answering service.
Visit our website at www.hoddereducation.co.uk.

© Karen Morrison 2008
First published in 2008 by
Hodder Education, an Hachette UK Company
Carmelite House, 50 Victoria Embankment
London EC4Y 0DZ

Impression number	14	13	12
Year	2022	2021	2020

Cover photo © Gary Vestal/Photographer's Choice/Getty Images
Illustrations by Robert Hichens Designs
Typeset in 12.5/15.5pt Garamond by Charon Tec Ltd (a Macmillan Company)
Printed in Dubai

A catalogue record for this title is available from the British Library.

ISBN 978 0 340 96603 7

Contents

Chapter 1   Doing science   1

Chapter 2   Characteristics of living things   17

Chapter 3   Cells and organ systems   27

Chapter 4   Classification and variation   41

Chapter 5   Understanding ecosystems   55

Chapter 6   Acids and bases   69

Chapter 7   Physical and chemical changes   79

Chapter 8   Materials and their properties   89

Chapter 9   The particle model   99

Chapter 10   Mixing and separating substances   111

Chapter 11   Forces   127

Chapter 12   Energy resources   137

Chapter 13   Electrical circuits   149

Chapter 14   The Earth in space   157

Glossary   175

Chapter 1 | Doing science

→ **Figure 1.1**
These pupils are doing science.

Science learning is active learning. This means that you will need to *do* things as you learn science. To succeed at science, you need to learn how to:

- work safely with science equipment
- gather and record information
- measure accurately and use the correct apparatus
- learn from diagrams
- read, understand and draw graphs
- find information and investigate how things work
- present what you have learned.

This chapter will teach you some of the skills that you need to do these things. The skills that you learn and practise as you work through the chapter will be re-used and developed further as you work through the rest of your science course.

Unit 1 Understanding the language of science

↑ **Figure 1.2**
The same word can have different meanings.

When you read science books, you will find words that you use in everyday life, such as the word 'space'. Some of these words have special meanings in science. For example, in science, 'space' means the stars, planets and other things beyond the Earth.

It is very important to use the correct scientific words to explain things in science. If you do this, your meaning is clear and you communicate your knowledge and understanding without confusion.

Learning science vocabulary

This year, you will learn some new science words. These words will help you increase your science vocabulary. The important science words are printed in **bold** the first time they appear in this book. Usually the meaning of the word is explained in the paragraph where it is found. For example:

A **flexible** material is one that bends easily when you apply a force to it. A **stiff** material does not bend easily.

Using a glossary

A glossary is like a mini-dictionary. The glossary at the end of this book is a list of all the words that are in bold in the chapters. The words are listed in alphabetical order to make it easy for you to find them. When you read a word that you don't understand, turn to the glossary and find the word in the list. Read the meaning that is given next to it to help you to understand it better.

Science and English

What do you do when you understand something but you don't know the English words to explain what you mean? This is a common problem for people who don't speak English as their first language. Many people know the words they want to use in their mother-tongue, but not in English. Here are some ways that you can use to find the English words and develop your vocabulary.

- Talk to friends who know your mother-tongue, ask them if they know the English words and get them to help you. If you are at home, your family may also be able to help.
- Ask your teacher to help you translate. This may not be possible if the teacher does not share your mother-tongue. But he or she will still be able to guide you to find the vocabulary you need.
- Use a bilingual or multilingual dictionary to find the words you need in English. Once you know the English words, you can check their meanings in the glossary.

Activity 1.1 **Using a science glossary**

Read the following passage about matter.

> All solids, liquids and gases around us are made of **matter**. Scientists believe that matter is made of tiny particles (pieces or bits) that clump together. You cannot see these particles, even with expensive equipment, but you can see the matter.
>
> The **particle model** of matter is a theory (a set of ideas) that explains how particles are arranged in solids, liquids and gases and how they behave in different conditions.

1 Use the glossary to find the meanings of the terms written in bold.

2 Use each term in a sentence to show its meaning clearly.

Unit 2 Safety in the science classroom

→ **Figure 1.3**
It is important to work safely in science.

Accidents can and do happen at school. It is important that you know what to do if someone hurts themself. Read the information below carefully. Your teacher will discuss this in more detail and tell you where to find the first aid equipment.

Cuts

Any injury involving flowing blood must be treated carefully to prevent the spread of diseases like TB and hepatitis and to prevent the slight risk of HIV infection. Wear rubber gloves or cover your hands with plastic bags before you touch blood. Help the injured person to clean the wound with water and disinfectant and then cover it with a waterproof dressing. Wash any equipment used in water and bleach, and burn any bloody waste such as used cotton wool.

Burns

You can be burned by hot water, steam, flames or chemicals. If a burn is red, but not blistered or burned through the skin, you should rinse it with cold water for 25 minutes and then cover the affected area with burn lotion and a sterile dressing. Blistered or burned-through burns need medical attention.

Eye injuries

Chemicals or small pieces of metal or wood can cause serious eye injuries. It is best to avoid these by wearing safety goggles, but if there *is* an accident, you have to act quickly. For chemicals in the eye, rinse the eye out with water. If there is a splinter in the eye, get the injured person to a clinic or doctor to have it removed carefully. Never try to remove objects from the eye yourself. Cover the eye with a cup and a dressing and get the person to a doctor.

 **Activity 1.2**   Safety rules

1 Work in small groups.
   a) Make a list of five accidents that might happen in a science classroom.
   b) Write down five ways in which your science classroom could be made safer.
   c) Compare your lists and safety suggestions with another group. Which safety issues do you think need most attention in your classroom?

2 Copy the table below into your book.
   a) Add four more safety rules, based on your decisions in question 1.
   b) Complete the second column by giving a reason why each rule is needed.

Safety rule	Why we need it
Don't run in the science classroom	
Tie back long hair	
Use tongs or gloves to carry hot equipment	
Never leave the Bunsen burner unattended	
Wipe up spilt chemicals immediately	
Wash equipment after using it	
Tell your teacher if you break something	
Listen carefully and follow instructions	
Use gloves	
Wear safety goggles	

3 Look at the pictures of pupils doing science below.
   a) In pairs, write down at least ten dangerous things that are happening.
   b) Refer to the list of safety rules you made in question 2. Which of these could help to prevent the dangers shown in the pictures?

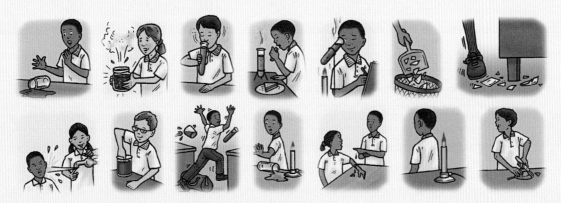

↑ **Figure 1.4** What are these pupils doing wrong?

Unit 3 Collecting and recording information

Scientists often work by asking a question, and then testing things to find the answer to the question. For example, scientists have asked the question: Can herbal medicines help to strengthen the immune system of HIV-positive people? In order to find the answer, they have developed a programme to test these medicines with volunteers. The tests are carefully planned and carried out to make sure they are fair and that the results are accurate. During testing, the scientists collect information or evidence to help them answer the question.

What can be tested?

It is not possible to find the answer to every question by testing. One of the first things that you will learn is how to identify a question that can be tested, from a set of possible questions. Testable questions usually ask what could be changed, observed or measured in an investigation. For example:

● What happens if I change the height of the ramp?
● Which type of paper absorbs the most water in a minute?
● How does the temperature of the classroom change during the day?

Questions that cannot be tested in the classroom are more general. For example:

● Which planets in our Solar System are made of rock?
● How are batteries made?
● What is the lifecycle of a frog?

You can still answer these questions, but you use different methods to find the answers, such as research and observation.

Fair testing

In a **fair test**, you change one factor (the **independent variable**) and measure its effect on another factor (the **dependent variable**). You *keep all other factors the same*, so you know that only the factor you are changing can have an effect on the dependent variable. This makes the test fair.

For example, you might want to find out which type of sugar dissolves fastest in water. Type of sugar is the independent variable in this test and time is the dependent variable. To investigate which type of sugar dissolves fastest in water, you would carry out tests using different types of sugar. You would keep the volume of water, the temperature of the water, the amount of sugar and the number of times you stirred the mixture the same for each test. You would measure how long it takes each type of sugar to dissolve in order to find the answer to your question.

Planning and carrying out a fair test

The outline below will help you to plan and carry out fair tests.

Our question:

Factor that we will change:

Factors that we will keep the same to make the test fair:

What we will measure:

How we will record our results:

What we will need to carry out our test:

 **Activity 1.3** **Planning and carrying out a simple test**

An art teacher asked the pupils in her class to tear strips of newspaper for a weaving project. Some pupils managed to tear straight strips, others struggled. The pupils wondered why this happened and decided to investigate whether the direction in which they tore the page (up and down, or across) made a difference to the straightness of the strip.

1 Work in small groups to plan and carry out a fair test to see whether the direction in which you tear newspaper into strips makes a difference to the straightness of the strip.
   a) Start by writing a testable question. For example:
      How does the _____ affect _____?
   b) Use the outline above to plan and record the results of your test.

2 What does your evidence suggest? Discuss this as a class.

Gathering evidence and recording findings

In Activity 1.3, you gathered evidence by tearing strips in different ways.

You then compared them to see whether the strips torn up-and-down the page were straighter than the strips torn across.

In other words, you made an **observation** by looking at the strips. Your observations allowed you to draw a **conclusion** – in this case, that the direction in which you tear affects or does not affect the straightness of the strip. This is just one way of gathering evidence in science.

Evidence from observation

In Activity 1.3, you observed with your eyes. However, in science, we use *all* our senses to make observations. For example, you could smell flowers to see which have a sweet smell and which smell unpleasant. You could taste different foods to see which ones are bitter and which are hot. You could feel different materials to see which are smooth and which are rough. You could listen to different sounds to see which is clearest from a distance.

Evidence from research

You can gather information to answer scientific questions from many different sources. Some of the most useful resources are books, magazines, newspapers and the internet.

Evidence from measuring

Measuring is an important method of collecting evidence in science. This year, you will measure length, area, mass, volume, temperature and time in order to draw conclusions about scientific happenings. Figure 1.5 shows you some of the ways in which you might do this.

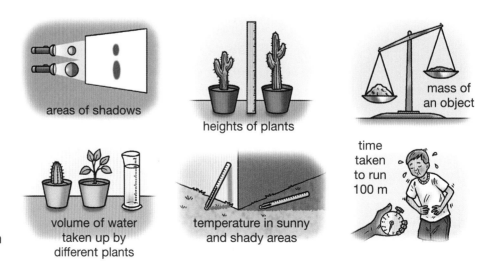

areas of shadows

heights of plants

mass of an object

volume of water taken up by different plants

temperature in sunny and shady areas

time taken to run 100 m

➡ **Figure 1.5** Measuring is an important skill in science.

Evidence from questionnaires, surveys and interviews

A **survey** is a way of gathering evidence. You can carry out a survey by asking people questions or by observing things (for example, counting the leaves on different plants).

When you ask questions, it is useful to write them down before you start your survey. You also need to know whether you want facts or opinions as evidence. For example, you might ask: Which type of batteries do you use at home? The answer will be a **fact**. You could also ask: Do you think that the government should supply free electricity to poor people? The answer will be an **opinion**.

Surveys and questionnaires have to be planned. The questions you ask should be clear and simple, and everyone you interview in the survey should be asked the same questions to make it fair. It might be easier for people to answer if they are given some answers to choose from. This is called 'multiple choice'.

Activity 1.4 **Deciding how to gather evidence**

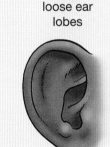

loose ear
lobes

joined ear
lobes

↑ **Figure 1.6**
Two different kinds
of ear lobes

Work in pairs.

1 Below are four questions a scientist might ask. Choose one of the questions. From the box underneath, select the method you would use to collect evidence to answer the question.

  1 People have either loose ear lobes or joined ear lobes (see Figure 1.6). Which type of ear lobe is most common in our class?

  2 Do families today have fewer children than families twenty years ago had?

  3 What do our classmates think of television and radio reports about environmental pollution in our country?

  4 How much water is wasted by dripping taps at our school?

  interview   measure   observe   survey

2 Plan and carry out your investigation – collect the evidence you need to answer the question. Record and keep your evidence. You will need it for Unit 6 (page 15).

Unit 4 Learning from diagrams

In science, you must be able to draw clear diagrams to show how you set up equipment in an experiment. Scientists use simple line diagrams rather than 3D drawings to show the equipment they use.

You will use these simple sketches. You will also learn to draw and interpret many other types of diagrams, including maps and graphs.

beaker

test tube

↑ **Figure 1.7**
How to draw
equipment

Cross sections and cut-away views

A cross section is a diagram which shows what an object would look like if you cut through it. The diagrams of the beaker and test tube in Figure 1.7 are cross sections. Cross sections are also used to show what the inside of something looks like.

A cut-away view shows you the outside of an object with parts of the inside also shown.

Flow diagrams

A flow diagram is a way of showing the ordered steps in a process. These can be in the form of pictures, or in the form of boxes with words, as in Figure 1.8.

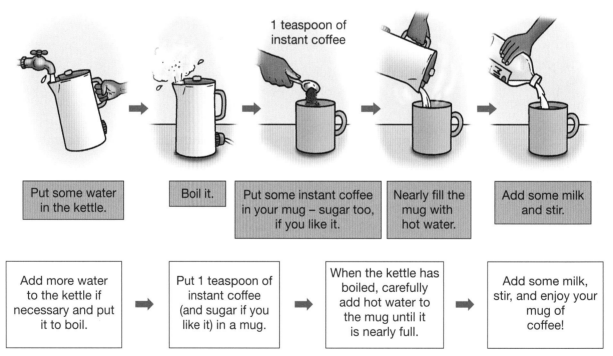

1 teaspoon of instant coffee

| Put some water in the kettle. | Boil it. | Put some instant coffee in your mug – sugar too, if you like it. | Nearly fill the mug with hot water. | Add some milk and stir. |

| Add more water to the kettle if necessary and put it to boil. | Put 1 teaspoon of instant coffee (and sugar if you like it) in a mug. | When the kettle has boiled, carefully add hot water to the mug until it is nearly full. | Add some milk, stir, and enjoy your mug of coffee! |

↑ **Figure 1.8** Flow diagrams give information in clear, easy-to-follow steps.

Headings, captions and labels

All scientific diagrams must have a heading or caption which tells you what the drawing shows. The parts of the diagram must also be clearly and correctly labelled. In some diagrams, particularly maps and graphs, information about the diagram is given in a key.

Rules for drawing in science

- Work in pencil.
- Use a ruler to draw straight lines.
- Draw cross sections of equipment.
- Don't draw lines across the tops of containers.
- Label all pieces of equipment. Label lines should never cross each other.
- Label any substances in the containers.
- Draw in the centre of the space and make drawings large enough to see them easily.

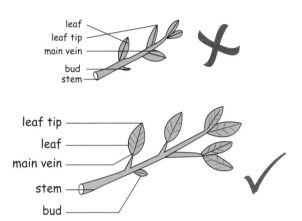

↑ **Figure 1.9** The lower diagram shows you the correct way to draw in science.

Activity 1.5 Working with diagrams

1 Draw a cross section to show what a boiled egg would look like if you cut it in half lengthways.

2 Look at Figure 1.10. Match the equipment shown to its name in the box.

measuring cylinder
test tube
round bottom flask
beaker
retort stand
Bunsen burner
watch glass

a)  b)   c)   d)

e)   f)   g)

↑ **Figure 1.10** Science equipment

3 Draw a simple flow diagram to show how you would safely light and use a Bunsen burner.

Unit 5 Graphs

When you collect evidence, it is sometimes useful to display it on a graph to make it easier to read and see patterns. You have already worked with graphs in mathematics lessons. This year, you will read, interpret and draw different graphs in science as well.

Different types of graphs

Type	Main features	Useful for
Bar graphs 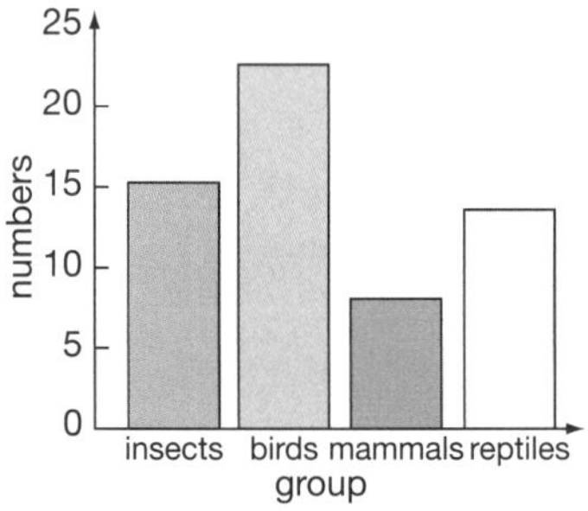 Figure 1.11 Bar graph showing the animal species found at school	Information shown as bars of equal width. Each axis is labelled to show what was measured. Scale on each axis gives measurements. Information given as exact amounts or percentages.	Comparing different sets of information which are not related to each other and which can be counted. This is called **discrete data**.
Line graphs 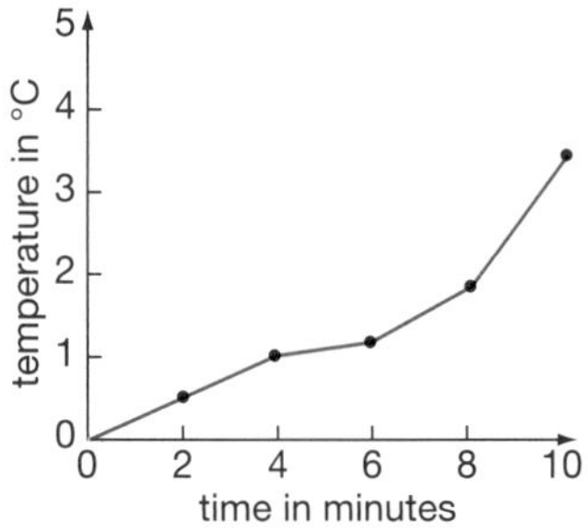 Figure 1.12 Line graph showing the temperature changes as ice melts	Points are plotted against both axes. Points are joined by a line. Both axes are labelled and give information.	Showing changes or patterns over time. Data which is measured is normally shown on a line graph. This is called **continuous data**.
Pie graphs 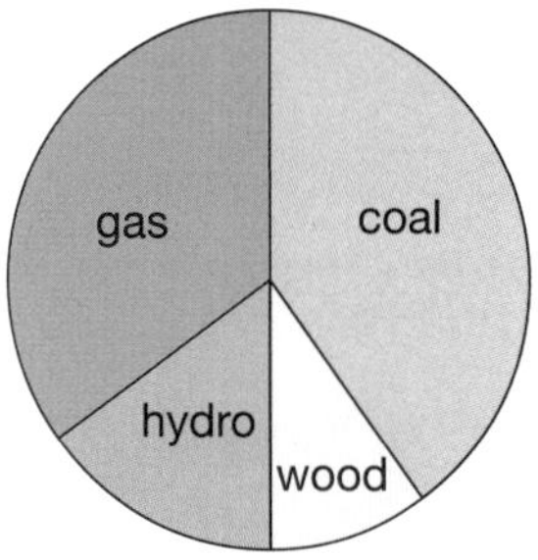 Figure 1.13 Pie graph showing the use of energy resources	Information shown as segments of the graph. Sectors labelled but figures often not given. Data needs to be converted to degrees to fit the circle.	Showing and comparing the size of different sections of a population or survey data.

Choosing the correct type of graph

You will have to draw bar graphs and line graphs in order to present evidence you have collected in a fair test. You will need to decide which type of graph is going to be most suitable. The information below can help you to do this.

- Choose a bar graph if the factor you change is expressed in words (for example, the *type* of sugar), but what you measure is expressed in numbers (for example, the *time* it takes to dissolve, in minutes).
- Choose a line graph if the factor you change is expressed in numbers (for example, the *mass* of a stone, in grams) and what you measure is also expressed in numbers (for example, the *length* of an elastic band, in centimetres).

Calculating percentages

Sometimes it is useful to compare quantities using percentages. Do you remember how to do this? Read the example carefully and make sure you understand how to change an amount to a percentage.

In a test, 12 out of 40 people preferred herbal tea to ordinary tea.

Therefore, the percentage of people who preferred herbal tea is

$$\frac{12}{40} \times \frac{100}{1} = 30\%$$

Activity 1.6 **Reading and drawing graphs**

1 Work in pairs. Talk about each of the graphs in the table on page 12. Discuss what is shown on each and why the person has chosen this particular type of graph to show the information.

2 Jane and Ishmael collected evidence during an experiment. They made this table of their measurements.

Temperature of water, in °C	20	30	40	50	60
Time it took for sugar to dissolve, in seconds	80	60	35	30	27

a) Show this information on a graph.
b) Write a few sentences explaining what this graph tells you.

Unit 6 Communicating your findings

There are many different ways of communicating your findings in science investigations to other people. You've already seen how you can use different types of diagrams and graphs to do this. However, these are just a few of the many methods you can use.

diagrams

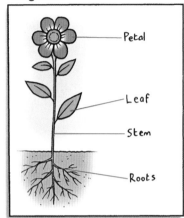

tables and graphs

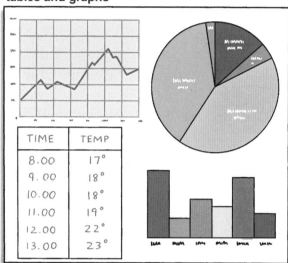

posters

maps

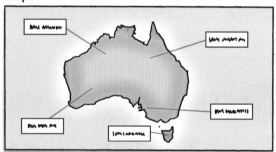

computer presentation or slide show

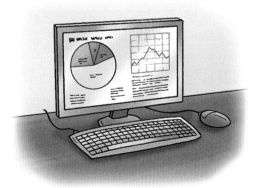

oral report

↑ **Figure 1.14** Different methods of presenting information

Activity 1.7 Presenting your findings

1 Look at the different ways of presenting findings on page 14. For each method of presentation, choose one type of information that could be presented in that way.

2 Use the evidence you collected in Activity 1.4, in Unit 3.
   a) Decide on the best method of presenting this evidence to the rest of your class.
   b) Present your evidence to the class.
   c) Ask the rest of the class to comment on your presentation and to suggest ways in which you could improve it.

3 Some pupils did an experiment to investigate plant growth. Their results are shown in Figure 1.15.

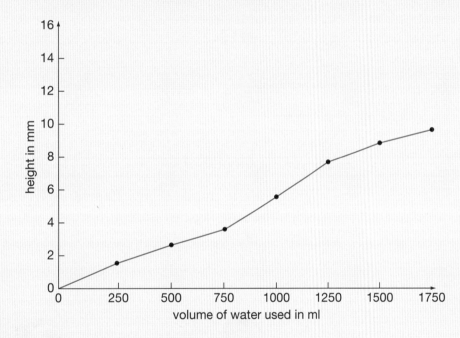

➡ **Figure 1.15**
Line graph showing the heights reached by plants given different volumes of water

   a) What do you think these pupils did to collect the evidence they used to draw this graph?
   b) What conclusions could they draw from their evidence?
   c) If you did a similar experiment with four different patches of grass, would you still choose a line graph to show your results? Explain your answer.

Chapter summary

In this chapter you have learned that:

- ☑ science involves active learning (doing things)
- ☑ a glossary is a useful tool for developing science vocabulary
- ☑ it is important to obey safety rules in the science classroom
- ☑ scientists ask questions that can be answered by carrying out investigations and fair tests
- ☑ testing involves gathering evidence in order to draw conclusions
- ☑ diagrams are important sources of information
- ☑ graphs allow us to show results and patterns quickly and clearly
- ☑ the ways in which we present our evidence and findings can make communicating our ideas much easier.

Revision questions

Work in pairs. Study the drawings in Figure 1.16 carefully, and answer these questions for each one.

1 What are the pupils doing? Explain fully, using correct scientific language when you need to.

2 What skills does each pupil need?

3 What safety measures should the pupils take?

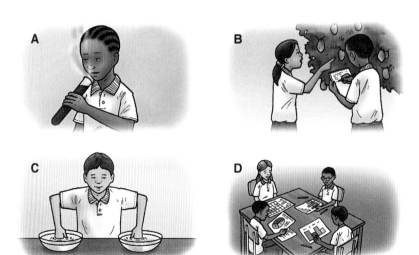

→ **Figure 1.16**
These pupils are all doing science.

Chapter 2

Characteristics of living things

⬆ **Figure 2.1** Which of these things are living?

If you look around you, you will see that there are many different things. Some things, such as people and animals, are alive. Other things, such as fallen leaves and dried beans, were once alive but are no longer living. Other things, such as rocks, water and plastics, are not alive and have never been alive.

In this chapter you are going to learn how to tell the difference between living and non-living things, and understand what makes a thing living or non-living. You will:

● use a set of characteristics to decide whether something is living or non-living

● look at each characteristic and observe it in action in living things.

Unit 1 Is it alive?

We can divide all the things on Earth into two groups – living things and non-living things. Animals and plants are living things. All living things have to do seven things to stay alive:

1 They take in oxygen (**respiration**).
2 They need food and water (**nutrition**).
3 They get rid of waste products (**excretion**).
4 They become bigger as they get older (**growth**).
5 They produce young (**reproduction**).
6 They can change position on their own (**movement**).
7 They sense things in the surroundings and react to them (**sensitivity**).

↓ **Figure 2.2**
The characteristics of living things

These seven life processes are characteristic of living things. We use these characteristics to decide whether something is living or non-living. If even one of these characteristics is missing, the thing is non-living.

Living things take in and get rid of gases, such as oxygen and carbon dioxide. They also excrete waste products such as sweat and urine.

Living things need food for energy. This animal is eating plants. The plants make their own food in their leaves.

Living things grow. These plants have grown from seeds. They are also growing towards the light because they are sensitive to light.

Living things produce young. These ducklings hatched from eggs. They will grow to become adult ducks.

Living things move. These animals swim in water.

Trees grow and produce seeds which can grow into new trees. The tree can also move its leaves so that they get as much light as possible.

Activity 2.1   **Deciding whether something is living or not**

↑ **Figure 2.3**

Look carefully at the things in Figure 2.3.

1 Draw a table like this one in your book.

Living things	Non-living things

2 Write the name of each thing in Figure 2.3 in the correct column in the table.

3 Choose one living thing. Explain how it shows each of the seven characteristics of living things.

4 Choose one non-living thing. Which characteristics is it missing?

Unit 2 Nutrition, respiration and excretion

Living things need food for energy, and so that they can grow. Getting and eating food is called **nutrition**.

Plant and animal nutrition

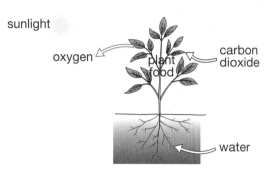

↑ **Figure 2.4** Plants make their own food by photosynthesis.

Plants make their own food. The green parts of plants contain a special chemical called **chlorophyll**. With this chemical, the plant is able to use sunlight to change carbon dioxide and water into sugars. This process is called **photosynthesis**.

Animals do not make their own food. They need to eat plants or other animals to get energy. Some animals, such as sheep and caterpillars, only eat plants. Some animals, such as humans, eat both plants and animals. Some animals, such as jackals and sharks, only eat other animals.

↑ **Figure 2.5**
Animals eat plants and other animals.

Respiration

All living things need energy. They get this energy from the food they eat. The process of getting energy from food is a chemical reaction called **respiration**. Respiration usually needs oxygen (but it can happen without oxygen).

In respiration, food combines with oxygen inside the bodies of plants and animals. During respiration, carbon dioxide and water are produced and energy is given off (released). We can show this as an equation:

food + oxygen → carbon dioxide + water + energy

Living things use the energy to move, grow and repair their bodies.

You should remember that respiration is not the same as breathing. When we breathe, oxygen moves into our bodies and carbon dioxide moves out of our bodies. You will learn more about this later.

Excretion

Excretion means getting rid of waste products that can harm the body. Many of these waste products are made by chemical reactions that take place inside the body. If the waste products are not excreted, they build up in the body and they can cause the animal or plant to die.

Animals excrete waste by breathing out carbon dioxide, by urinating and by sweating. You can do a simple experiment with limewater to show that the air you breathe out has carbon dioxide in it. Limewater turns white and milky or cloudy when carbon dioxide is present.

Plants cannot breathe, urinate or sweat. Plants excrete oxygen, which is a waste product of photosynthesis, through tiny pores in their leaves.

Experiment 2.1

↑ **Figure 2.6**
How to set up your experiment

Testing for carbon dioxide

You will need:
● a straw ● a stirring spoon ● two dishes with limewater in them

Method

1 Place one dish of limewater near a window where it will get fresh air. Stir the limewater with the spoon so that it mixes with the air. What happens to the limewater?

2 Put one end of the straw into the other dish of limewater. Put the other end of the straw in your mouth. Take a breath in through your nose and slowly blow it out through the straw. What happens to the limewater?

The limewater in the first dish does not go milky because fresh air does not have large amounts of carbon dioxide in it. This is the air we breathe in. The limewater in the other dish turns white and milky because the air we breathe out has high levels of carbon dioxide in it.

Activity 2.2   Investigating local plants and animals

1 Make a list of five plants and five animals that live in your area.

2 How does each plant and animal get the food it needs?

3 How does each plant and animal get rid of waste products?

4 What is the main difference between plant nutrition and animal nutrition?

Unit 3 Growth and reproduction

Growth

All living things grow. When things grow they get bigger. Living things can grow longer. Plant stems and roots become longer as the plant grows. Humans get taller and our limbs get longer as we grow. Living things can also grow wider and develop new parts. The trunks of trees get wider and new branches develop as the tree grows. In animals, the body gets wider and some animals develop horns or wings as they grow.

Most plants grow all of their lives. Most of the growth in a plant takes place in the roots and stems.

Humans and other animals stop growing when they reach their full size. We say they have reached **maturity**.

As plants and animals grow, they also develop. **Development** means becoming more complicated. For example, tadpoles do not have legs. These develop as the tadpole grows into a mature frog. Young human boys do not have beards and young girls do not have breasts. These features develop as we grow and develop into mature adults.

Animals grow in different ways. The process of growing from birth, to maturity, to death is called a **lifecycle**. Figure 2.7 shows you two different animal lifecycles.

↑ **Figure 2.7** Butterflies and chickens have different stages of growth and development in their lifecycles.

Reproduction

You have seen that birth and death are part of all lifecycles. All plants and animals have to make new plants and animals just like themselves. Making new versions of yourself is called **reproduction**. For example:

- dogs give birth to young dogs (puppies)
- mosquitoes lay eggs which hatch and develop into new mosquitoes
- adult palm trees grow coconuts which fall from the tree and grow into new trees
- mushrooms make spores which grow into new mushrooms.

↑ **Figure 2.8** Plants like these reproduce by making seeds which grow into new plants.

Some living things can reproduce from parts of themselves. For example:

- strawberry plants and spider plants grow new plantlets at the ends of runners produced by the parent plant
- ginger plants grow from new buds on the roots of the parent plant, and new potatoes can grow from the 'eyes' on the parent potato
- simple animals such as the amoeba reproduce by splitting themselves into two – each part of the parent becomes a new amoeba.

↑ **Figure 2.9** Animals reproduce by laying eggs or giving birth to young.

Activity 2.3 Understanding growth and reproduction

1 How can you tell if something has grown?

2 Why do living things need to reproduce?

3 Copy and complete this table to describe how these plants and animals grow and reproduce.

Living thing	Signs of growth	How it reproduces
mosquito		
chicken		
rabbit		
human		
maize plant		
palm tree		
thorny cactus plant		

Unit 4 **Sensitivity and movement**

Sensitivity and sense organs

Animals can sense changes in their surroundings using their sense organs. The sense organs are the eyes, ears, nose, mouth and skin. We say that animals are **sensitive** to their surroundings. This is an important characteristic because it allows wild animals to sense danger and react to keep themselves alive.

Woodlice are insects which live in cool, damp, dark surroundings. Figure 2.10 shows how these insects respond when conditions in their environment change.

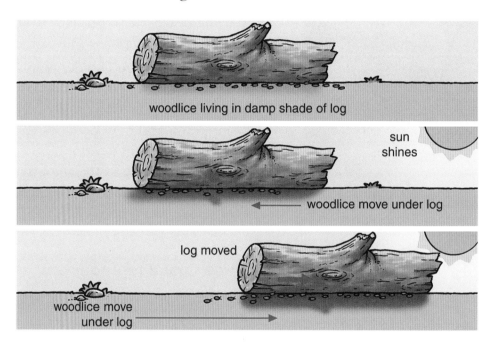

woodlice living in damp shade of log

sun shines

woodlice move under log

log moved

woodlice move under log

→ **Figure 2.10**
The woodlice react to light and heat by moving away.

↑ **Figure 2.11**
Insects like this stink bug have long sensitive antennae.

Some animals have long, sensitive whiskers. Cats, for example, have whiskers which help them sense conditions in their surroundings. Some insects have long antennae which they use to feel changes in their surroundings (Figure 2.11).

Plants do not have sense organs like animals do, but they are still sensitive to their surroundings. Plant roots are sensitive to gravity – that is why they will grow down into the soil even if you plant the seed upside down. The leaves and stems are sensitive to light, so plants will grow upwards or sideways to get more light. Some plants, like the mimosa tree in Figure 2.12, are sensitive to touch. When they sense touch, the leaves close up. This is to prevent them being eaten by grazing animals.

Movement

↑ **Figure 2.12**
The leaves of the mimosa close when they are touched.

Animals can move parts of their bodies. You are able to move your arms, legs, fingers, tongue, eyelids and nose using your muscles. But animals can also move around from place to place by walking, swimming or flying. They move around to find food, to find shelter and to escape from danger.

Some movements in the body are unconscious. This means they happen without you thinking about them or controlling them. For example, your heart beats and pumps blood around your body. You do not control this movement – your brain does it automatically. The muscles in your intestines move automatically to push food through the digestive system.

Plants cannot move around from place to place without help. But they can move parts of themselves. For example, the roots can move down through the soil, and the stems and leaves can move towards light.

Activity 2.4 | **Describing sense organs and movement**

Look at the animal in Figure 2.13 carefully.

➜ **Figure 2.13**
A locust

1 Describe how the animal senses things in its environment.

2 This animal uses its antennae to touch the plants or ground in front of it. How do you think the information it gets helps it to stay alive?

3 Give three ways in which this animal can move from place to place.

4 Explain why plants cannot move from place to place by themselves.

Chapter summary

✓ Living things share certain characteristics:
- They feed (nutrition).
- They respire to get energy from food.
- They get rid of body wastes by excretion.
- They grow and develop as they mature.
- They reproduce to produce new versions of themselves.
- They are sensitive to their surroundings.
- They can move parts of themselves and may move from place to place.

✓ All plants and animals are living things. They all share these characteristics although they may show them in different ways.

Revision questions

For each behaviour in the list below, write down one or more correct characteristics of living things from the box underneath.

- smelling food
- escaping from danger
- producing fruit
- hearing a noise
- giving birth to a baby
- lifting a heavy object
- chewing grass
- getting taller

- using oxygen to release energy
- sweating
- eating an apple
- feeling tired because you have run fast
- laying eggs
- urinating
- forming new plantlets
- bending towards light

nutrition   respiration   excretion   growth   reproduction   sensitivity   movement

Chapter 3 — Cells and organ systems

➜ Figure 3.1
This scientist is
studying cells
using a
microscope.

Scientists like to know how living things work. To find out more about living things and how they function, scientists study cells. In this chapter, you will learn about microscopes and how they have helped scientists to learn more about cells. You will also learn about the work that cells do in plants and animals and how they group together to form tissues and organs.

To understand what cells are and how they work, you need to be able to:

● understand how a microscope works
● discuss how microscopes have helped scientists learn more about cells
● draw diagrams of plant and animal cells to show what they look like under a microscope
● compare plant and animal cells
● give examples of cells that perform special jobs in plants and animals
● explain how cells are organised to form tissues and organs
● identify the main organ systems found in humans
● list the functions of different organ systems.

Unit 1 Using a microscope to look at cells

A microscope is a piece of equipment that makes objects look bigger. When the object looks bigger, you can see more detail in it. The microscope in Figure 3.2 uses two lenses to **magnify** objects (make them look bigger). This type of microscope is called a light microscope. The parts are labelled and you can read what each part does.

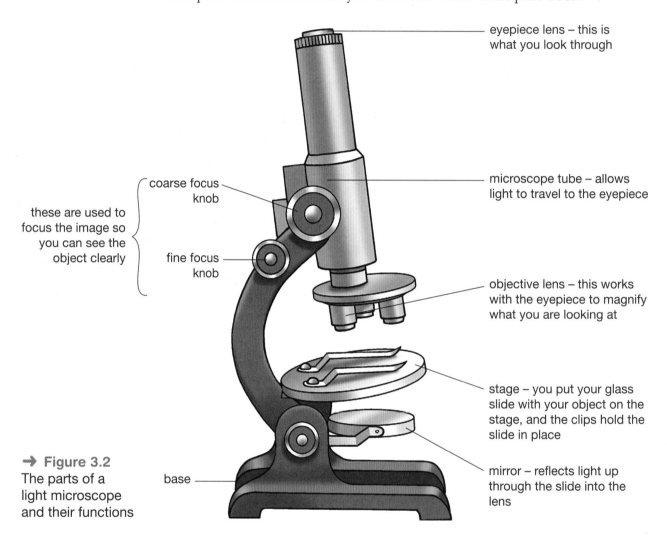

eyepiece lens – this is what you look through

microscope tube – allows light to travel to the eyepiece

coarse focus knob

these are used to focus the image so you can see the object clearly

fine focus knob

objective lens – this works with the eyepiece to magnify what you are looking at

stage – you put your glass slide with your object on the stage, and the clips hold the slide in place

mirror – reflects light up through the slide into the lens

base

→ Figure 3.2
The parts of a light microscope and their functions

Before the 1600s, no one had ever seen a cell. But, in 1665, a scientist called Robert Hooke used a very simple microscope to look at some cork. The microscope allowed Hooke to magnify the cork 30 times. This means that each detail in the cork looked 30 times bigger than it really was.

Hooke was able to see that the cork was made up of small box-shaped structures. He called these cells, because they looked like small rooms in a monastery or prison.

Figure 3.3 shows you what cork cells look like under a microscope.

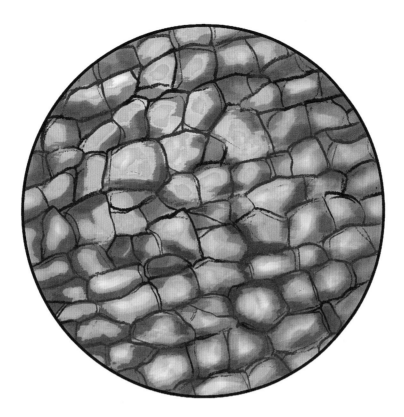

→ **Figure 3.3**
This is what Robert Hooke saw when he looked at cork cells under a microscope.

In 1675, another scientist called Anton van Leeuwenhoek used a better microscope to look at a drop of pond water. He was able to see tiny living things in the water.

As microscopes improved, scientists were able to look at plants and animals in more detail. This helped them to discover that:

- all living things are made of cells
- cells are building blocks that join together to make complicated structures (such as flowers, stems, eyes, hearts and brains)
- cells can only be made by other living cells.

 **Activity 3.1**   **Answering questions about microscopes**

1 What is a microscope and what is it used for?

2 Why can't we see cells without a microscope?

3 Who was the first person to see cells under a microscope? Describe what he saw.

4 What does it mean if a microscope magnifies an object 300 times?

5 What did scientists find out about cells by studying objects with microscopes?

Unit 2 Studying and drawing cells

50×

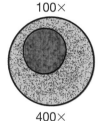

100×

400×

↑ **Figure 3.4**
The field of view gets smaller as you increase the magnification.

Magnification

Before you look at cells using a microscope, you need to know about **magnification**. In a microscope, this depends on the strength of the eyepiece and the objective lenses. The strength of each lens is normally written on it. For example, you may have an eyepiece lens with a magnification of 5× and an objective lens with a magnification of 30×. To find the total magnification, you multiply these two amounts: $5 \times 30 = 150\times$. This means the object you are looking at is magnified 150 times, so it looks 150 times bigger than it really is.

Drawing what you see

When you look through a microscope, you see a circle of light with some, or all, of the object in it. This circle of light is called your **field of view**. If you increase the magnification, your field of view gets smaller because you are looking at the object in more detail. This happens in real life too – hold out your hand and look at it. You can see your whole hand. If you move your hand closer to your eyes, you can see a smaller part of your hand, but you can see more detail in the part you can see. Figure 3.4 shows you a cell at three different magnifications.

When you draw what you see through the microscope, it is important to give an idea of scale. Follow these steps:

1 Start by drawing a circle to represent your field of view.
2 Then draw your object so that it takes up the same proportion of the space in your drawing as it does when you look at it.
3 Write down the magnification you are using and what you are looking at.

Soil grains
400×

↑ **Figure 3.5** A completed drawing of soil grains seen under a microscope.

Experiment 3.1

Looking at onion cells

You will need:
● a microscope ● a small piece of onion ● a glass slide ● a cover slip
● a straw or dropping pipette ● water ● a mounted needle

Method
1 Peel off a very thin piece of onion.
2 Put the piece of onion onto your glass slide.
3 Add a drop of water onto the slide using the straw or pipette.

4 Carefully lower the cover slip onto the onion using the mounted needle. Try not to trap any air bubbles under the cover slip.

5 Place your slide on the stage of the microscope. Hold it in place with the clips.

6 Look through the eyepiece. Move the mirror so that you get a clear circle of light.

7 Look at the microscope from the side. Carefully and slowly turn the coarse focus knob till the objective lens is just above your slide.

8 Look through the eyepiece and turn the fine focus knob until the onion cells come into focus.

9 Draw what you see at two different magnifications.

Experiment 3.2

Looking at human cheek cells

You will need:
- a microscope ● a clean cotton bud ● a glass slide ● a cover slip
- iodine solution to stain the cells ● a straw or dropping pipette
- water ● a mounted needle

Method

1 Gently scrape the inside of your cheek with the cotton bud.

2 Rub the cotton bud onto the slide.

3 Add a drop of water and a drop of iodine solution, and then lower the cover slip on top.

4 Examine your cheek cells through the microscope on low, medium and high magnification.

5 Draw what you see.

Safety note

To avoid any chance of infection, use a clean cotton bud for each person. Do not share them. Throw the cotton buds into the bin when you have finished. Put the slides with their cover slips into warm soapy water when you have finished with them, to clean them.

 **Activity 3.2** **Comparing plant and animal cells**

Look at your drawings from Experiments 3.1 and 3.2.
What differences can you see between the plant and the animal cells that you examined?

Unit 3 Comparing plant and animal cells

All living things are made up of cells. Simple animals, like the amoeba, and bacteria only have one cell. But most plants and animals contain millions or billions of cells.

Plant cells and animal cells contain some of the same parts and they sometimes do the same things. But plant cells have some parts not found in animal cells because plant cells have to do different jobs from animal cells.

Figure 3.6 shows you a typical plant cell and a typical animal cell.

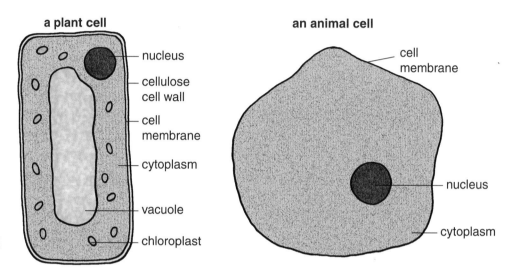

a plant cell — nucleus, cellulose cell wall, cell membrane, cytoplasm, vacuole, chloroplast

an animal cell — cell membrane, nucleus, cytoplasm

➜ **Figure 3.6**
The structure of plant and animal cells

If you look at Figure 3.6, you will see that both plant cells and animal cells have three basic parts:

- **Cell membrane** – this is a layer of 'skin' around the cell. The membrane gives the cell its shape. It also allows substances to move into the cell and out of the cell.
- **Cytoplasm** – this a watery jelly-like substance that fills up the cell. Many chemical reactions take place in the cytoplasm. These are important for cell functioning.
- **Nucleus** – this is a darker area in the cell. We can say the nucleus is like the 'brain' of the cell. The nucleus controls all cell functions. It also contains all the chemical 'instructions' necessary to make new cells.

Plant cells have some extra parts that are not found in animal cells:

- **Cell wall** – this is a strong layer of tough material (called cellulose) that supports the cell and makes it keep its shape. The cell wall is found outside the cell membrane.
- **Vacuole** – this is a large space in the centre of the cell. The vacuole contains a mixture of water and dissolved sugars and salts called the **cell sap**. The cell sap helps the cell to keep its shape.

- **Chloroplasts** – these are small structures found in the cytoplasm. They contain the chemical **chlorophyll**. They give the plant a green colour and are important because they allow plants to make their own food using sunlight.

Activity 3.3   Differences between plant cells and animal cells

1 Copy the table and make a tick (yes) or a cross (no) to show whether each part is found in plant cells and animal cells. The first row has been done for you.

Part of the cell	Plant cells	Animal cells
cell wall	✓	✗
cell membrane		
cytoplasm		
nucleus		
vacuole		
chloroplasts		

2 Copy this table into your book and complete it to summarise the differences between plant and animal cells.

Plant cells	Animal cells
Have cell walls made from cellulose	
Contain chloroplasts	
Have a vacuole containing cell sap	
Often shaped like a box	
Nucleus is often at the edge of the cell	

3 Why do cells have a cell membrane?

4 What are the functions of chloroplasts?

5 Why is the nucleus an important part of a cell?

Unit 4 Specialised cells

Cells come in different shapes and sizes. The shape and size of a cell depends on what **function** (job) it has in the plant or animal where it is found.

Specialised animal cells

Most plants and animals contain millions of cells. The bodies of humans and other animals are made up of many different types of cells. Each type of cell has a special job to do. We say that cells are **specialised** when they are adapted to suit the job they do. Figure 3.7 shows you some of the different types of cells found in humans and other mammals. In Unit 5 you will learn more about how these cells work together in the body.

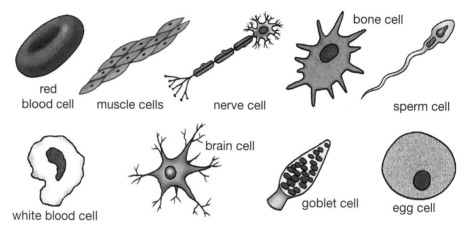

→ Figure 3.7 Some of the specialised cells found in humans (not to scale)

Specialised plant cells

Plant cells also have special jobs to do. Some cells make food, some cells absorb water and minerals from the soil, some cells carry water up the stem to the leaves, and other cells protect the plant. Figure 3.8 shows you some of the different types of cells found in plants.

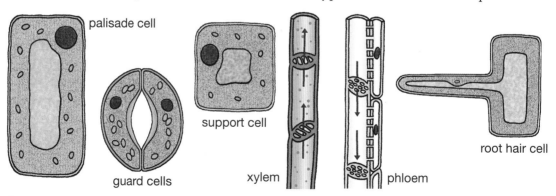

↑ Figure 3.8
Some of the specialised cells found in plants (not to scale)

Activity 3.4   **Matching cells to their functions**

1 Write down the names of five different types of specialised animal cells. Next to each type of cell, write down its function in the body.

2 Match each of the cells in column A with a function from column B.

Column A – cells	Column B – functions
palisade cells	protect the openings on leaves
root hair cells	contain chloroplasts
xylem cells	transport water through the plant
support cells	absorb water from the soil
guard cells	help the leaf keep its shape

3 What types of cells are shown in Figure 3.9?

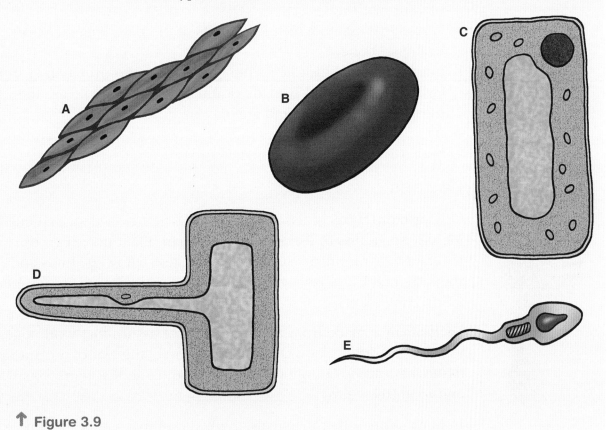

↑ **Figure 3.9**

Unit 5 Cells are organised

Tissues

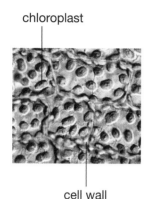

chloroplast

cell wall

↑ **Figure 3.10**
Leaf tissue in a
moss plant

Cells are normally found in groups. A group of the same type of cells which do the same job is called a **tissue**. Tissues are found in both plants and animals. Figure 3.10 shows you how leaf cells fit together to form leaf tissue in a moss plant. The leaf tissue allows the plant to make its own food by photosynthesis.

Plants also have special tissues that allow them to transport water and food, and to make seeds and fruits.

Animals, including humans, have many different types of tissues. Smooth muscle cells form muscle tissue, nerve cells form nerve tissue and skin cells form skin tissue. Even blood is a type of tissue. Body tissue can be placed in one of four groups according to its function. These groups are:

- epithelial tissue – covering tissue (for example, skin and the lining of your intestines)
- connective tissue, which joins parts of the body together (for example, bone)
- muscle tissue, which helps the body perform movements (for example, your heart is made of muscle tissue, and so are the muscles in your arms and legs)
- nerve tissue, which carries messages between the body and the brain (for example, your spinal cord).

Organs

bud

flower

leaf

roots

↑ **Figure 3.11**
The organs of a
flowering plant

One or more types of tissue may group together to do a particular job. These groups of tissues form an **organ**. Plant tissues group together to form organs such as leaves, stems, flowers, buds and roots. Figure 3.11 shows you where these main organs are found on a flowering plant.

Each organ is made from different kinds of tissues that work together to carry out different functions. For example, the leaf of a plant has tissue that carries out photosynthesis, and tissue that transports water and food (the veins).

In humans and other mammals, the heart, stomach, lungs, brain, skin and kidneys are all organs. Each of these organs is made from different kinds of tissues which work together to help the organ do its job.

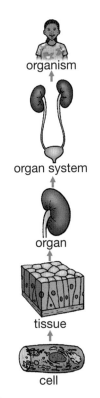

organism

organ system

organ

tissue

cell

Figure 3.12
How cells are related to organ systems

Organ systems

Organs cannot work on their own. They have to work together with other organs to carry out all the jobs needed to keep the plant or animal alive. For example, plant roots can take up water and minerals but they cannot transport these to the rest of the plant without the stem and leaves. In animals, the heart cannot move blood around the body without the veins and arteries.

When a group of organs work together to carry out certain jobs, they are called an **organ system**. For example, humans have a digestive system which is made up of the throat, stomach and intestines. Humans also have a circulatory system made up of the heart, veins and arteries, and blood. Each organ system carries out particular functions in the body. You will learn more about the important organ systems in the human body in Unit 6.

Figure 3.12 shows you how cells are related to organ systems.

Many living things have more than one organ system working together to keep the organism alive. **Organism** is the correct biological name for a living thing.

Activity 3.5 · Completing sentences

Use the key words in the box to complete the sentences below.

circulatory
organism
system
specialised
digestive
muscle
cells
root
nerve
organ
nervous

1 Cells are adapted to do different jobs. We say they are _____.

2 A tissue is made up of a group of specialised _____.

3 _____ tissue is an example of a tissue found in plants.

4 Two examples of tissues found in animals are _____ tissue and _____ tissue.

5 An _____ is a group of different tissues which work together.

6 When organs work together to carry out certain functions they are called an organ _____.

7 The human body has many different organ systems. These include the _____ system, the _____ and the _____ system.

8 _____ is another name for a living thing with different organ systems.

Unit 6 Human organ systems

Can you remember the seven characteristics of living organisms? These are nutrition, respiration, excretion, growth, reproduction, movement and sensitivity.

In humans, our organ systems help us to carry out these important life processes. The table shows which organ systems help us to carry out each of these life processes.

Life process (characteristic)	Organ systems
nutrition	digestive system, circulatory system
respiration	respiratory system, circulatory system
excretion	excretory system, circulatory system
growth	endocrine system
reproduction	reproductive system, endocrine system
movement	muscle system, skeletal system
sensitivity	nervous system

The diagrams in Figure 3.13 show you where these important systems are found in the human body. The diagrams also show you the main organs in each of these systems. Study these diagrams carefully to learn more about each system.

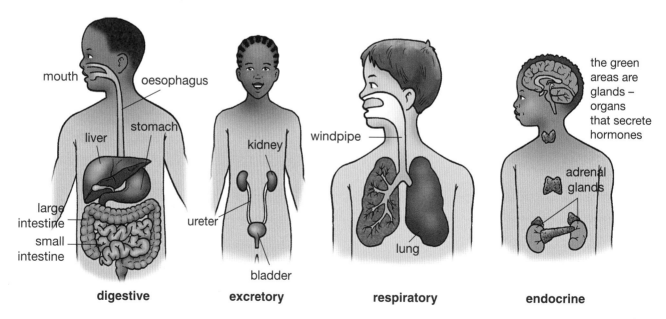

↑ **Figure 3.13** The main organ systems of the human body

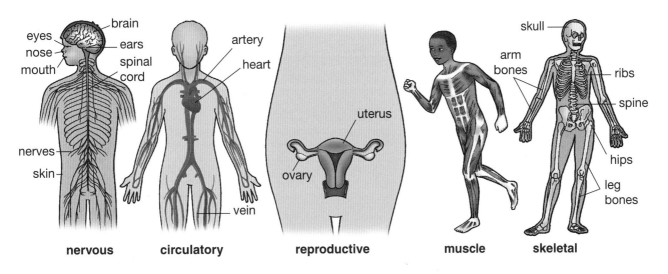

nervous   circulatory   reproductive   muscle   skeletal

⬆ **Figure 3.13** The main organ systems of the human body (*continued*)

Activity 3.6 Matching organs to systems and functions

1 List five organs found in the digestive system.

2 Name the sense organs which feed information into the nervous system.

3 What is the main job of the circulatory system?

4 Which organ system controls growth in humans?

5 Match each set of organs listed below with an organ system from the box.

- heart, veins and arteries
- lungs and windpipe
- adrenal glands
- brain and spinal cord
- biceps and triceps
- eyes, ears, nose, mouth and skin
- kidneys and bladder
- ovaries and uterus
- thigh bone, ribs and spine
- stomach and intestines

digestive system
circulatory system
excretory system
respiratory system
endocrine system
reproductive system
muscle system
skeletal system
nervous system

Chapter summary

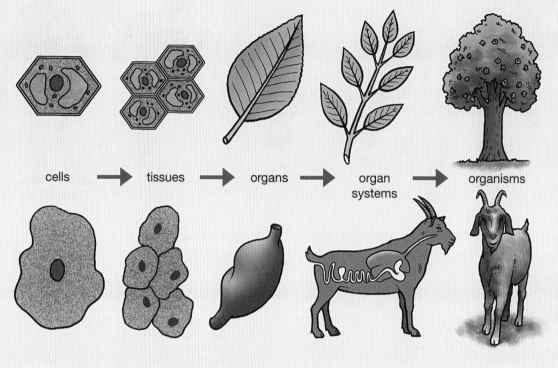

cells → tissues → organs → organ systems → organisms

↑ **Figure 3.14** Building blocks of organisms

Revision questions

1 Draw and label a plant cell and an animal cell to show how they are different.

2 Explain what we mean when we talk about cell specialisation.

3 What is a tissue? Give an example of a tissue found in plants and a tissue found in animals.

4 What do we call groups of tissues that work together?

5 Name five organs found in flowering plants.

6 Give five examples of organs found in the human body.

7 What is an organ system?

8 Name one organ system found in humans. Which organs are found in this system? What is the main job of this system?

Classification and variation

→ **Figure 4.1** The vegetables on this stall have been arranged in groups.

Grouping vegetables fruit on a stall makes it easier for us to find what we want to buy. There are many different ways of grouping the vegetables. They could be grouped by size, shape, colour or type.

When scientists group things, we call it **classification**. Organisms are normally classified by the things they have in common.

In this chapter, you will learn more about classification. You will learn how to use a classification key to find out which group an organism belongs to. You will also learn how all the plants and animals in the world can be classified into major groups. Then you will study some examples to find out more about these groups. Lastly, you will learn about variation and the small differences between organisms that belong to the same group.

As you learn about classification and variation, you will:

- describe how and why scientists classify organisms
- understand what is meant by a species
- use similarities and differences to classify plants and animals into major groups
- use and draw up different kinds of classification keys
- name the major kingdoms used to classify living organisms
- give examples of organisms belonging to groups in the plant and animal kingdoms
- explain and investigate variation in humans and other species.

Unit 1 Organising living things

You have already learned how to group things into 'living' and 'non-living'. There are over 1.85 million named organisms on Earth. Every year, new organisms are discovered and this number increases. No one knows how many types of organisms there are on Earth, but scientists estimate that there could be between 5 million and 50 million different types.

Scientists group organisms together using **characteristics**. This process of grouping things together is called **classification**. Organisms are grouped together because they have things in common or because they are similar in some way. For example, all animals that have bones, gills and scales, and live in water, are classified as fish.

There are many different reasons why scientists classify organisms. Arranging things in groups creates a sense of order.

↑ **Figure 4.2** Clothes are organised into sections in shops so that similar items are close together.

Think about a large clothing store. If the clothing was just placed anywhere, it would be difficult to find things. Usually the store groups all the women's clothes together and all the men's clothes together. The women's and men's clothes are then divided up so that the trousers, blouses, sleepwear and underwear are in separate sections. This makes it easier to find what you want when you go shopping.

Placing things into groups makes it easier to describe things. You can use one word to describe many different members of a group. For example, we can use the words 'insects', 'flowers' and 'fish' to describe the items in each of the three groups shown in Figure 4.3.

These are all insects. These are all flowers. These are all fish.

↑ **Figure 4.3**

When scientists classify a living thing, they use a special scientific name for it. These scientific names are accepted and used all over the world so that scientists in one country know which living things other scientists are talking about.

Each unique group of living things is called a **species**. The members of a species can successfully reproduce with each other. Every species is given a two-part scientific name. The first part is the **genus** and the second part is the species. The genus always has a capital letter but the species does not, and the words are normally written in italics. These scientific names for species are important because one species may have many common names and this can lead to confusion. For example, one species of juicy fruit, *Annona squamosa*, is known by many different local names in different countries (Figure 4.4).

Classification helps scientists to work out how organisms are related to other organisms using similarities and differences.

When we classify organisms we learn about the number of living things in different environments. This helps us to learn more about the relationships between living things in the environment. When we know which living things are threatened, we can take action to protect and conserve them for the future.

You should remember that classification systems are based on what we know about living things at the moment. When we learn more about living things, the ways in which we group and classify them may change.

↑ **Figure 4.4**
Annona squamosa is known by many different local names – cherimoya, custard apple, sweetsop and qu a na, among others.

Activity 4.1 **Classifying organisms**

↑ **Figure 4.5**

1 Put the plants in Figure 4.5 into two groups.
2 Regroup the plants so that you have four groups.
3 What characteristics did you use to make two groups?
4 What characteristics did you use to make four groups?
5 Could you use different characteristics to make groups? Discuss this.
6 Could the method you used to classify the plants lead to any confusion? Explain why or why not.

Unit 2 Using classification keys

Living things can be classified using a **key**. In a key, there are normally two choices for each characteristic. This type of key is called a dichotomous key. 'Dichotomous' means divided into two parts.

In a classification key, you start with one big group. For example, in the key in Figure 4.6 we have started with trees. You then divide the group into smaller groups using one important characteristic. Each group is then divided again into two smaller groups, until the key names all the things that you have to classify. To use the key you follow the branches that match the characteristics of the thing you are trying to identify.

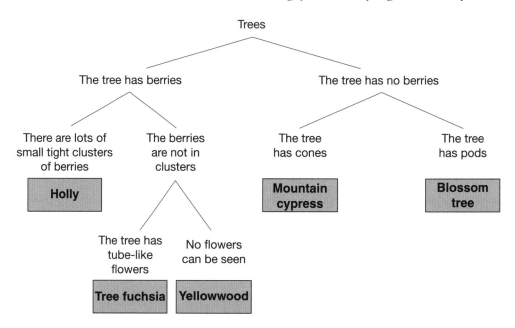

➡ Figure 4.6
A branching
dichotomous key to
classify tree species

Keys can be drawn as a branching diagram as in Figure 4.6. But they can also be written in words with instructions rather than branches to follow.

The key for sorting the trees in Figure 4.6 can be written in words like this.

1	The tree has berries. The tree has no berries.	Go to number **2**. Go to number **4**.
2	There are lots of small tight clusters of berries. The berries are not in clusters.	**Holly** Go to number **3**.
3	The tree has tube-like flowers. No flowers can be seen.	**Tree Fuchsia** **Yellowwood**
4	The tree has cones. The tree has pods.	**Mountain cypress** **Blossom tree**

Activity 4.2   Using a key to name living things

The diagram shows branches from five trees.

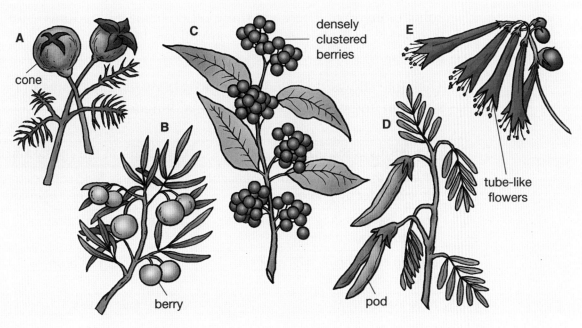

↑ **Figure 4.7**
Branches from
five different trees

1 Identify trees A, B and C using the key in Figure 4.6.

2 Use the word key on page 44 to find the names of trees D and E.

3 Make a key to classify the large cats in Figure 4.8. Use the labels to help you. You can use either a branching key or a word key.

➡ **Figure 4.8**
Five large cats

Unit 3 Plants and animals

Most living things can be placed into one of five big groups, called **kingdoms**. The two main kingdoms are the plant kingdom and the animal kingdom. There are three other smaller kingdoms – fungi, Protoctista and Monera – but you will not learn about these this year.

The table shows the main differences between organisms in the plant kingdom and organisms in the animal kingdom.

Organisms in the plant kingdom	Organisms in the animal kingdom
produce their own food from sunlight, air and water by photosynthesis	need organic food sources to survive; cannot produce their own food
do not have feeding structures such as mouth and intestines	have feeding structures
contain chlorophyll, which allows them to make food	do not contain chlorophyll
have leaves	do not have leaves
have roots	do not have roots
do not move around from place to place	move around
do not have nerves and muscles	have nerves and muscles
do not have sensory organs such as eyes and nose	have sensory organs such as eyes and nose or other receptors

Plant groups

The plant kingdom can be divided into smaller groups. There are four main group: mosses, ferns, conifers and flowering plants.

Mosses are small plants that live in damp places. They can't survive in hot, dry places, as they lose water through their thin leaves and can't transport water through the plant. Mosses produce spores, not seeds.

→ Figure 4.9
Moss

↑ **Figure 4.10** Ferns

Ferns are found naturally in cool, damp places. They are bigger than mosses. Ferns have strong roots, stems and leaves. The leaves hold water and the plant can transport water, so they can survive in places that are not damp. Ferns also produce spores, not seeds.

↑ **Figure 4.11** Cones on a pine tree

Conifers are plants that bear cones. They normally have thin, needle-shaped leaves, which they keep all year. They have a water transport system and waterproof leaves. The plants produce seeds, which are formed inside cones. Conifers range from small shrubs to giant trees.

↑ **Figure 4.12** Flowering plants come in all shapes and sizes.

Flowering plants produce flowers. The flowers produce seeds, inside fruits. Flowering plants have a water transport system and usually have broad, waterproof leaves. Most plants we see fall into this group – from tiny daisies to flowering grasses to massive trees.

 **Activity 4.3** **Classifying plants**

1 Make a list of at least ten different plants that you know.
2 How could you sort the plants on your list into smaller groups?
3 Discuss the characteristics that you would use to do this.

Animal groups

Like plants, the animal kingdom can be divided into smaller groups. The animals that we can see without a microscope can be divided into two groups according to whether or not they have a backbone. We call animals with backbones **vertebrates** and animals without backbones **invertebrates**.

Invertebrates

Invertebrates come in all shapes and sizes, and many of them have soft bodies. Some have shells outside their bodies (e.g. snails) or a hard exoskeleton (e.g. beetles). Others have shells inside their bodies (e.g. cuttlefish). Invertebrates do not have bones or a skeleton.

There are more invertebrate species than vertebrates, but the vertebrates are more noticeable because they are normally bigger. They can grow larger because their skeletons allow them to support a larger body.

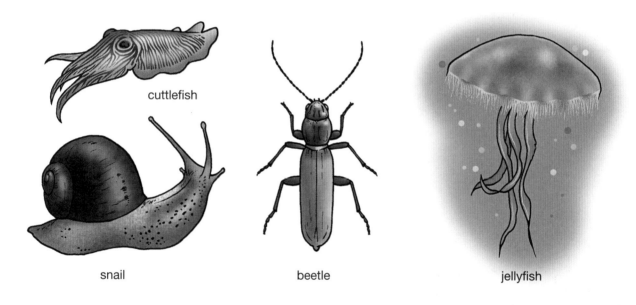

cuttlefish

snail

beetle

jellyfish

↑ **Figure 4.13** Different kinds of invertebrates

Vertebrates

All vertebrates have a hard skeleton inside their bodies. This gives support to the body and protects the internal organs. It also allows the animal to move around. The backbone of a vertebrate is made up of lots of smaller bones called **vertebrae**. If you run your finger down your own backbone you will feel bumps – these are your vertebrae.

human   bird   rabbit   fish

⬆ **Figure 4.14** All vertebrates have an internal skeleton made up of hard bones.

Activity 4.4   Classifying using a table

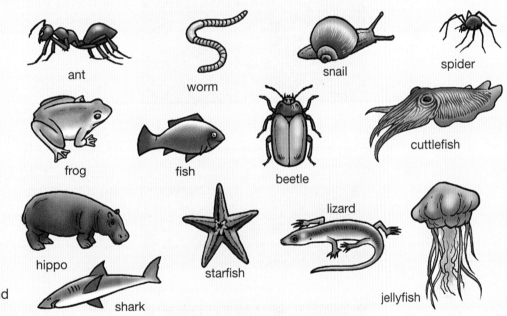

ant

worm

snail

spider

frog

fish

beetle

cuttlefish

➡ **Figure 4.15**
This is a mixture
of vertebrates and
invertebrates.

hippo

shark

starfish

lizard

jellyfish

1 Make two columns in your notebook, one for vertebrates, the other for invertebrates.

2 Write the names of the animals in Figure 4.15 in the correct columns.

3 Add two more animals to each column.

4 What can you say about the sizes of the animals in the invertebrate column?

Unit 4 Classifying vertebrates

There are five main groups of vertebrates: fish, amphibians, reptiles, birds and mammals.

↑ Figure 4.16 Different kinds of fish.

Fish live in water. They breathe through gills. Their bodies are covered with scales and they use fins to swim.

↑ Figure 4.17 A frog is an amphibian.

Amphibians are animals that can breathe on land and in water, like frogs. They have smooth, damp skin and they breathe using lungs as well as through their skin. They lay their eggs in water.

↑ Figure 4.18 A crocodile is a reptile.

Reptiles breathe air using lungs. They have dry, scaly skin and they lay eggs on dry land.

↑ Figure 4.19 Flamingos are birds.

Birds are warm-blooded animals with two legs, two wings and feathered bodies. Most of them can fly. They have beaks and they lay eggs with hard shells.

↑ Figure 4.20 Cows are mammals.

Mammals are warm-blooded animals that have hair on their bodies. Their young develop inside the mother and are born alive (not inside an egg shell). The mother makes milk in her body to feed the young.

Classifying people

All people on Earth belong to the species *Homo sapiens*.

'Homo' means human. All members of this genus walk on two legs and have a round jaw with similar teeth.

'Sapiens' means wise. All members of this species have a very large frontal part to their brain. They can learn things and they develop culturally. **Culture** is the knowledge that different generations develop and pass on to the following generations.

People are able to think about ideas, use words to talk and imagine things they can't see. This makes them different from other animals.

 **Activity 4.5** **Classifying vertebrates**

1 Read carefully each description below, and then decide which group each animal belongs to.
   a) My skin is dry and scaly and I lay my eggs in sand.
   b) I feed my babies on milk that I produce. My body is covered in hair.
   c) My body is always warm and it is covered in feathers. I lay my eggs in a nest.
   d) I need water to lay my eggs. My skin is smooth and I can breathe through it.
   e) I live in water and breathe through my gills. I have a scaly body.

2 Human beings are vertebrates. Which sub-group of vertebrates do we belong to? Why?

Unit 5 Variation

↑ Figure 4.21 People are all the same species, but they are all different.

↑ Figure 4.22 Each of these giraffes is different from all the others although they all look the same.

↑ Figure 4.23 Each apple on these trees will be slightly different from all the other apples.

There are almost seven billion people on Earth. All these people belong to the same species, but they are also individuals who are **unique** (one of a kind). Every animal in every species on Earth is unique as well. Even trees and plants of the same species are different from each other in small ways.

The differences between members of the same species are called **variations**. In humans, hair and eye colour, mass, height and facial features are just a few of the **characteristics** that differ from person to person.

Parents and their children

When you look at brothers and sisters, you can often tell that they come from the same family because they have similar **features**. The similarities between parents and their **offspring** suggest that some characteristics are **inherited**. In other words, the characteristics are passed naturally from parent to child.

But not all characteristics are inherited. You might be born with eyes that look exactly like your mother's eyes, but features such as height and mass are affected by diet and environment as much as by inheritance.

Variation and the environment

A survey of variation amongst 200 pupils at three different schools looked at height and blood groups. The results are shown in Figures 4.24 and 4.25.

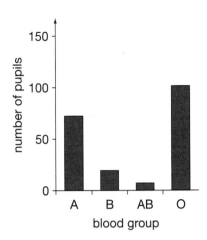

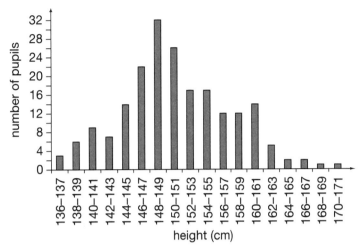

↑ **Figure 4.24** Blood type is inherited. It cannot be affected by the environment.

↑ **Figure 4.25** You might inherit the characteristics for tallness, but if you do not eat properly, you will not grow to your full height. Height is affected by the environment.

Activity 4.6 **Investigating variation**

1 Indira drew up this diagram showing some characteristics of her family over three generations.

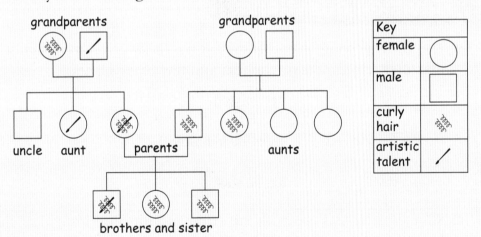

→ **Figure 4.26**

When Indira's friend Sami looked at her family tree diagram, he said that curly hair and artistic talent are inherited characteristics.
a) Do you agree with what he says?
b) Give two reasons for your answer.

2 Do you think being good at subjects like science or mathematics is an inherited characteristic? Give reasons for your answer.

3 Study the graph in Figure 4.25, which shows pupils' heights.
a) How many pupils were taller than 160 cm?
b) Why do you think there is such a variation in height among the pupils? Give two reasons.

Chapter summary

☑ We can classify organisms by putting them into groups.

☑ A species is a group of organisms that share the same characteristics, and that can reproduce with each other to produce young of the same species.

☑ Branching keys and word keys can be used to classify organisms.

☑ There are two main groups of organisms: the plant kingdom and the animal kingdom.

☑ There are four main groups in the plant kingdom: mosses, ferns, conifers and flowering plants.

☑ Animals can be divided into vertebrates and invertebrates.

☑ There are five main groups of vertebrates: fish, amphibians, reptiles, birds and mammals.

☑ The individual members of a species show variation. This means they are all slightly different from each other.

☑ Variation can be inherited but the environment can also lead to variation in a species.

Revision questions

1 Use the following characteristics in the order given to design a key which classifies ferns, mosses, conifers and flowering plants.

has seeds	has no seeds
seeds are in cones	seeds are in flowers
has a stem	has no stem

2 Why do scientists use body structure and function instead of size, colour or behaviour when they are classifying animals?

3 Copy and complete the following sentences to show what you know about the plant kingdom.

_____ are small plants that are found in damp places. They produce spores. _____ have a water transport system, but they also produce spores. Conifers often have small _____ leaves. They produce seeds, which are carried in _____. The _____ plants produce _____, which grow in fruits.

4 State one difference between:
   a) fish and amphibians   b) fish and birds   c) fish and reptiles
   d) fish and mammals   e) reptiles and mammals.

Understanding ecosystems

→ Figure 5.1
Where would you find living organisms in this place?

Living things are found in many different places and conditions. In this chapter, you are going to learn more about the places in which animals live. You will also find out how animals and plants manage to survive in different places and different conditions. Lastly, you will learn how living organisms are linked to each other through feeding relationships and food chains.

To understand and learn about living things and where they live, you will:

- learn and use correct scientific words to describe organisms and where they live
- list the things that organisms need in their habitats
- learn how organisms are adapted for survival
- study examples of animal adaptations
- investigate how plants are adapted to different environments
- describe different types of feeding behaviour
- interpret and draw simple food chains.

Unit 1 Understanding your environment

Your surroundings are your **environment**. Everything around you is part of your environment. If you are in a classroom environment, then you are probably surrounded by living and non-living things such as desks, books, a teacher, other pupils, windows, doors, plants, air and germs.

Organisms are not found just anywhere in the environment. We do not find fish living in trees or birds living under water. The type of place where an organism lives is called its **habitat**.

Living organisms need particular things in order to survive. They will only live in habitats where they can find the things they need to survive. For example, tropical fish need warm salty water, so they are found in habitats like the Indian Ocean where the water is warm and salty. Tropical fish are not found in the cold, icy waters near the poles because those waters do not provide a suitable habitat for them.

Some of the things that organisms need in their habitats are:

- a source of food
- water
- air
- light
- a suitable temperature.

Living things do not live alone. One habitat may be home to different groups. Each group of organisms of the same species living in one habitat is called a **population**.

When two or more populations are found living in the same habitat, they form a **community**. For example, the starfish, anemones, seaweeds and fish in the rock pool in Figure 5.2 are a community.

A community of living things together with the non-living things in the environment (the air, water and soil) form an **ecosystem**.

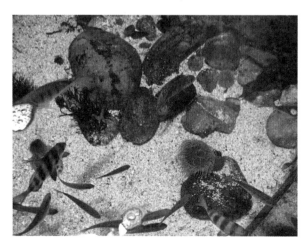

➡ Figure 5.2
This aquatic habitat is home to populations of starfish, anemones, seaweeds and fish.

There are many different ecosystems on Earth. Some ecosystems are found on land, others are found in water. Land ecosystems are called **terrestrial** ecosystems. Fresh and saltwater ecosystems are called **aquatic** ecosystems.

The pictures in Figure 5.3 show you four different environments with different conditions.

↑ Figure 5.3 What conditions would you expect to find in each of these places?

 **Describing environments**

1 Write a sentence to explain what the following words mean:
   a) environment   b) habitat   c) population
   d) community   e) ecosystem

2 Study the environments shown in the photographs in Figure 5.3.
   a) What conditions would you expect to find in each one?
   b) Make a list of the plants and a list of the animals that you would expect to find in each of these environments.
   c) Why would you not find exactly the same plants and animals in each place?

Unit 2 Adapting to different environments

Organisms are suited to the habitats that they live in. We say that they are **adapted** to their habitats. An **adaptation** is a special characteristic of an organism that helps it obtain food and water, to move from place to place, to protect itself from its enemies, to find shelter or to reproduce.

Adaptations may be behavioural (what an animal does). For example, oxpeckers live near cattle or wild animals so that they can feed on ticks on the animals.

Adaptations may be structural. This means that part of the animal's body is adapted for a special purpose. For example, the leafy sea dragon is a type of fish which has leafy flaps in its body to help it hide in seaweed. Mole rats have sharp claws and large front teeth to help them dig in soil.

Adaptations may also affect an animal's body functions. For example, snails seal themselves inside their shells during dry conditions. Polar bears hibernate during the winter to stay warm and save energy.

Figure 5.4 shows you a tree frog in its natural habitat in the tropical rainforest. Read the information round the photograph carefully to see how this small frog is adapted to its habitat.

feet shaped like suction cups for gripping leaves

red eyes to scare off predators

strong back legs for jumping

hiding in shady trees under leaves helps to keep the skin moist

↑ Figure 5.4 The red-eyed tree frog is adapted to suit its natural habitat in the rainforests.

Plants are also adapted to suit their habitats. Water lilies like those in Figure 5.5 are adapted to survive in water. They have large flat leaves that float on top of the water and their stems are hollow with large air pockets to help the plants stay upright in water. Their roots drift down into the water and their seeds are carried from place to place by water movements.

Cactuses like the one in Figure 5.6 are well adapted to dry conditions. They have thick, spongy stems that can store lots of water, thin spiky leaves to prevent water loss and deep grooves in their stems to allow them to expand and fill up with water whenever it rains.

↑ Figure 5.5 Water lilies are adapted to live in water.

↑ Figure 5.6 Cactuses like these are well adapted to dry conditions.

 **Activity 5.2**

Discussing adaptations

1 What is an adaptation? Give an example of a plant adaptation and an animal adaptation.

2 Choose a plant that grows naturally in your area. Draw the plant and label your drawing to show how the plant is adapted to the conditions in its environment.

Unit 3 Animal adaptations

Animal species adapt to their habitats for many different reasons. Some of these reasons are described here with examples.

Protection and defence

Some animals protect themselves from their natural enemies by blending into their surroundings, or camouflaging themselves. Some fish are adapted so that they look like the seaweed they live in. The bright green colour of tree frogs makes it easy for them to hide in leaves, and many lizards are a speckled brown colour so that they blend in with the rocks they live on.

→ Figure 5.7 This caterpillar has spots that look like eyes on its rear end. It also has special red horns that shoot out when it is threatened and give off a bad smell.

Other animals have developed defensive adaptations. The poison arrow frog gives off a deadly poison when it is attacked. The frog is also brightly coloured to put off attackers. Butterflies and caterpillars have false 'eyes' to scare predators. Sea anemones have stinging cells in tentacles to protect themselves from being eaten.

Coping with different conditions

Animals may need to survive in very hot, very cold, very wet or very dry conditions. There are many adaptations to help them do this.

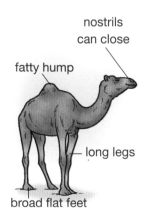

nostrils can close

fatty hump

long legs

broad flat feet

↑ Figure 5.8
Camels are very well adapted to live in hot, dry conditions.

Small desert mice and rats have hairy feet so they do not get burned on the hot sand while others have large ears that allow them to lose heat easily. Animals such as whales and seals, which live in cold conditions, often have a layer of fat below their skin to keep their bodies warm.

Lions in the swampy areas of Botswana, and tigers in the forests of India, have adapted their behaviour and often swim to cross areas of water. Camels store water in their humps so that they can survive in dry conditions, and beetles in the desert catch dew on their back legs so that they can drink it. Some rainforest frogs lay eggs surrounded by a thick, wet jelly to allow their tadpoles to develop without the need for a pond.

↑ Figure 5.9 Geckos have small suction caps on their feet which allow them to walk up walls and upside down on ceilings.

Movement

Animals need to be able to move easily from place to place. Structures such as flippers, legs or wings that are adapted to their living conditions help them to do this.

Food

Many animals have adaptations to help them catch food more easily. Many birds have adapted specially shaped beaks to help them catch and eat food. Figure 5.11 shows you some of the different types of beaks found on wading birds.

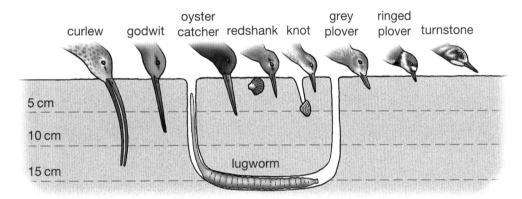

↑ Figure 5.10 Each bird's beak is adapted to suit the food it catches and eats.

Barnacles, which live on rocks in the tidal zone of a beach, have feathery feet which they use to pull food into the openings on their shells.

Breathing air

Some animals live in water, but they need to breathe air to get oxygen for respiration. Fish have gills that allow them to breathe oxygen from the sea water. Whales breathe through a special blowhole on the top of their heads. African lungfish have lungs outside their bodies which they use to breathe air when their water supply dries up.

 **Activity 5.3** | **Design your own animal**

1 Design an animal that can live in very cold, dry conditions such as those found at the top of a high mountain.
2 Make a list of the conditions that the animal will have to live with.
3 Draw your animal and show how it is adapted to its habitat.

Unit 4 **Plant adaptations**

If you look at an atlas, you will see that different parts of the world have different climate conditions and vegetation (plant) types. The region in which a particular type of climate and vegetation is found is called a **biome**. The table describes the conditions found in seven different biomes and shows how the plants in each biome are adapted for survival.

Biome	Conditions found there	How plants are adapted
desert	<ul><li>dry and often hot</li><li>low average rainfall</li><li>lots of direct sunlight</li><li>sandy or rocky soils</li><li>strong dry winds</li><li>big temperature difference between day and night</li></ul>	Figure 5.11 The spiny cactus has spines instead of leaves, and stores water in its stem. The hairy cactus has light-coloured hairs that help shade the plant. The Aptenia plant has a waxy coating on its leaves to reduce water loss.
grasslands	<ul><li>hot summers and cold winters</li><li>unreliable rainfall with frequent drought</li><li>rich soils</li></ul>	Figure 5.12 Soft stems enable prairie grasses to bend in the wind, and narrow leaves minimise water loss in the exposed, windy conditions of grasslands. Many grasses are wind-pollinated.

Biome	Conditions found there	How plants are adapted
tropical rainforest	<ul><li>hot with very high rainfall</li><li>flooding and soil erosion are common</li><li>poor soils</li><li>thick canopy prevents sunlight reaching the forest floor</li></ul>	   Figure 5.13 Drip-tips on leaves help to shed excess water, while buttress roots help support plants in the shallow soil. Some plants grow on other plants and don't have their own roots, so they collect rainwater in a central reservoir and absorb it through hairs on their leaves.
temperate rainforest	<ul><li>fairly constant temperature with mild winters and cool summers</li><li>high rainfall</li><li>poor soils</li></ul>	  Figure 5.14 Trees can grow very tall in this moist environment. Epiphytes live on other plants to reach the sunlight.

Biome	Conditions found there	How plants are adapted
deciduous forest	<ul><li>four clear seasons</li><li>temperature ranges from hot in summer to below freezing in winter</li><li>high rainfall</li><li>fairly fertile topsoils</li></ul>	  Figure 5.15 Many trees have thick bark to protect against the cold winters in the temperate deciduous forest. Broad leaves can capture a lot of sunlight for a tree, but in the autumn, deciduous trees shed their leaves to reduce water loss. 
coniferous (boreal) forests	<ul><li>cold winters and warm summers</li><li>some parts have a permanently frozen layer of soil called permafrost</li><li>water cannot drain well so ground is often swampy</li><li>precipitation is often in the form of snow</li><li>soils are acidic and fairly infertile</li></ul>	  Figure 5.16 Needle-like leaves help to reduce water loss and aid in the shedding of snow. The shape of many conifer trees also helps shed heavy snow, to save the branches from breaking.
tundra (cold desert)	<ul><li>cold all year with only a short cool summer and very long winter</li><li>permafrost layer</li><li>poor water drainage</li><li>little precipitation, usually as snow or ice</li><li>very long days in summer, very short days in winter</li></ul>	 Figure 5.17 Tundra plants are low-growing. Some grow in a clump which helps conserve heat.

Aquatic habitats

Plants that live in aquatic habitats also show adaptations. Some of these are:

- underwater leaves and stems, which are flexible so that they do not get broken by water currents
- air spaces in their stems so that they float and hold the plant up in the water
- roots and root hairs are often small or absent because the plant only needs roots to hold it in place, not to take up water
- leaves may float on top of the water so they are exposed to sunlight
- plants may produce seeds that can float (such as the coconut).

 **Activity 5.4** | **Linking adaptations to functions**

1 Copy and complete the table by filling in the functions of each plant adaptation listed.

Adaptation	Function
fleshy leaves	
needle-like leaves	
large broad leaves	
flexible stems and branches	
thick bark	
buttress roots	
waxy coating on leaves	
long, deep roots	

2 Draw a labelled sketch of a desert plant showing how it is adapted to the conditions in its environment.

Unit 5 Feeding relationships

All living organisms need food. Most organisms get the food they need from their habitat.

Plants are the only organisms that can make their own food, so they are called **producers**.

Animals are called **consumers** because they consume (eat) plants or other animals.

Animals such as sheep and cattle, which eat only plants, are called **herbivores**.

Animals such as lions, eagles and spiders, which eat only other animals, are called **carnivores**.

Animals such as bears, monkeys and humans, which eat both plants and animals, are called **omnivores**.

When a plant or animal dies, its remains are broken down by bacteria and fungi. These small organisms are called **decomposers**.

Food chains

Living things in every community are linked together in **food chains**, as shown in Figure 5.18.

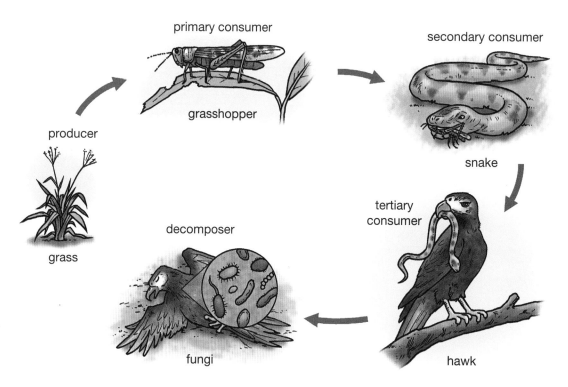

↑ **Figure 5.18** All food chains follow a similar pattern.

Food chains are normally shown as simple flow diagrams. The arrows show the direction in which energy is passed along the food chain from one living thing to another. For example:

grass → locust → snake → eagle → bacteria

All food chains start with plants and they usually end with bacteria or fungi (decomposers).

Activity 5.5   **Interpreting food chains**

1   Represent each of the three food chains in Figure 5.19 as a flow diagram using words and arrows only.

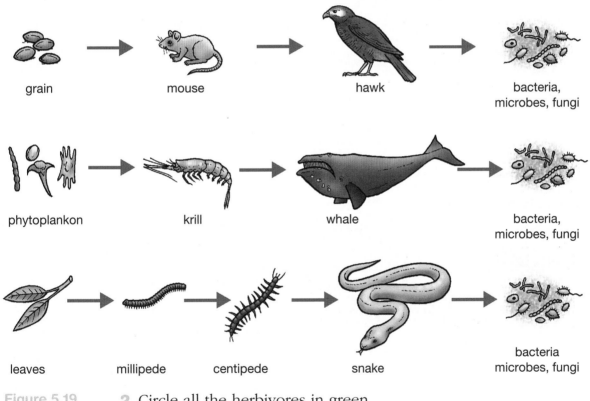

grain   mouse   hawk   bacteria, microbes, fungi

phytoplankon   krill   whale   bacteria, microbes, fungi

leaves   millipede   centipede   snake   bacteria microbes, fungi

↑ Figure 5.19

2 Circle all the herbivores in green.

3 Circle all the carnivores in red.

4 Circle all the omnivores in yellow.

5 List the producers, consumers and decomposers in the food chains.

Chapter summary

✓ The environment is another word for the living and non-living things in your surroundings.

✓ The place where a species lives is called its habitat.

✓ A group of organisms belonging to the same species that live in a habitat are called a population.

✓ The different populations that live together in a habitat are called a community.

✓ Living things are adapted to suit the environments they live in.

✓ Adaptations help animals to protect themselves, to survive in hot, cold, wet or dry conditions, to catch food, to move from place to place and to get oxygen for respiration.

✓ Plants are adapted to the climate and conditions in their environment. Adaptations help plants to save water, get maximum sunlight and successfully reproduce.

✓ Food chains are diagrams that show feeding relationships. Food chains always begin with plants, which are called producers because they produce food, using sunlight. All other organisms are called consumers.

✓ Herbivores eat plants, carnivores eat meat and omnivores eat plants and meat.

✓ The remains of dead animals and plants are broken down by decomposers.

Revision questions

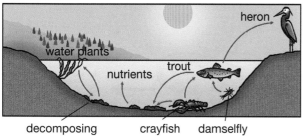

⬆ **Figure 5.20**
A pond ecosystem

1 Look at the diagram in Figure 5.20. In this ecosystem, name:
   a) two habitats   b) three populations
   c) one producer.

2 What does a heron eat?
   What do we call this type of animal?

3 Where does the initial source of energy for all food chains in this ecosystem come from?

4 Draw one food chain that is likely to exist in this ecosystem.

5 Describe the natural habitats in which you are likely to find the following organisms:
   a) tadpoles   b) conifers   c) camels   d) seagulls
   e) antelope   f) polar bears   g) anemones   h) cactuses

6 Why is it an advantage for a lizard to be the same colour as a rock?

Acids and bases

⬆ Figure 6.1 Many of the substances that we use in our daily lives are made from chemicals. The chemicals in the substances can make them acid or alkaline.

In this chapter you will learn more about acids, bases (alkalis) and neutral substances. You will use indicators to find out whether substances are acids or bases. Then you will use the pH scale to measure levels of acidity and alkalinity in substances. Knowing about acids and bases allows scientists to neutralise substances. You will learn about neutralisation and its uses in medicine, at home and in industry.

The work that you do in this chapter will allow you to:

- list the properties of acids and bases
- classify everyday substances as acids, bases or neutral
- explain how an indicator works
- make use of indicators to find whether a substance is acid or alkaline
- use and understand the pH scale
- explain how acids and bases can be neutralised
- give examples to show how neutralisation is used in daily life and in industry.

Unit 1 Acids

→ Figure 6.2
These things all
contain acid.

Acids are an important group of chemicals which have similar properties.

Some acids are weak. Most of these are acids that are found naturally in foods or in animals. Examples of weak acids are:

Toxic

The substance is
poisonous. Do not
inhale or swallow it.

- citric acid found in citrus fruit (oranges and lemons) and tomatoes
- ascorbic acid (also called vitamin C) found in fruit and vegetables
- tartaric acid found in grapes
- malic acid found in green apples
- lactic acid found in sour milk and yoghurt
- acetic acid (also called ethanoic acid) found in vinegar
- carbonic acid found in fizzy drinks like lemonade and cola
- formic acid (also called methanoic acid) found in stinging plants like nettles and in the saliva of some biting ants.

Irrltant

The substance will
make your skin
red and blistered if it
touches you.

Weak acids make food taste sour and they can make your teeth feel rough.

Other acids are strong and they can be dangerous. Sulphuric acid is a strong acid found inside car batteries. If this acid drops onto your clothes it will burn a hole in them. If you touch it, it will burn your skin. We say these acids are **corrosive** because they wear down or weaken other substances.

Corrosive

Corrosive acids will
seriously burn your
skin, and corrode
metals and other
materials.

Hydrochloric acid is another strong acid, which is used in industry to clean bricks and metals. This acid is also found in your stomach where it helps break down your food.

↑ Figure 6.3
Warning labels
found on acids

When you work with acids in the laboratory you need to handle them very carefully. Most acids that you use in the school laboratory have warning signs on them. Make sure that you know what the three signs shown in Figure 6.3 mean.

Experiment

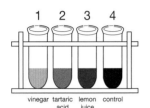

1 2 3 4

vinegar tartaric lemon control
        acid juice

test tubes filled
with black tea

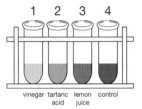

1 2 3 4

vinegar tartaric lemon control
        acid juice

test tubes filled
with beetroot water

↑ Figure 6.4
How to set up your
test tubes

Investigating the effects of acids on dyes

Aim
To investigate what weak acids do to the colour of different dyes.

You will need:
- eight test tubes ● a dropper or small funnel ● strong black tea
- ● lemon juice ● red water from boiled beetroot ● vinegar
- ● tartaric acid (cream of tartar) mixed with water

Method
Half fill four test tubes with black tea. Half fill the other four with the red beetroot water.
Add a few drops of vinegar to the first test tube in each set.
Add a few drops of tartaric acid to the second test tube in each set.
Add a few drops of lemon juice to the third test tube in each set.
Observe what happens in each of the test tubes.

Recording your results
Copy this table and complete it to record the results of your experiment.

Acid	Effects on black tea	Effects on beetroot water
vinegar		
tartaric acid		
lemon juice		

Questions
1 Explain why you did not add anything to the last test tube in each set.
2 What can you conclude from this experiment?

Activity 6.1

Identifying acids

1 a) Find five other examples of substances in your home that contain acids.
   b) Write down the name of each substance and the name of the acid(s) it contains.

2 If you eat too many oranges, the skin at the side of your mouth may become red and sore. Can you explain why this happens?

Unit 2 Bases

Bases are another group of corrosive chemicals. When they are mixed with water they feel slimy or soapy to the touch. Strong bases can burn your skin and you should never taste or touch them.

Most bases are solid and do not dissolve in water. Bases which do dissolve in water are called **alkalis**. So, for example, copper oxide is a base. It does not dissolve in water. Sodium hydroxide (caustic soda) is a base. But it does dissolve in water, so we call it an alkali.

When bases are mixed with acids they react strongly and cause the acid to lose acidity and the base to become less basic. This is why scientists sometimes say that bases are chemically opposite to acids.

Figure 6.5 shows you some of the bases and alkalis that you may have seen at home.

→ Figure 6.5
These substances are all bases (or alkalis).

Naming chemicals

All acids contain hydrogen. This means that they have the chemical symbol H in their shorthand chemical name. Hydrochloric acid contains hydrogen and chlorine. In chemistry, this acid would be called hydrogen chloride and written in shorthand as HCl. Sulphuric acid contains hydrogen, sulphur and oxygen and would be written in shorthand as H_2SO_4. When acids react with other substances they give up some or all of their hydrogen. You will learn more about this when you deal with chemical reactions.

Many bases have the word 'hydroxide' in their chemical names. The shorthand for hydroxide is OH. Caustic soda, which is used in drain cleaners, has the chemical name sodium hydroxide. This can be written in shorthand as NaOH. Slaked lime, which is added to acid soils to make them less acid, has the chemical name calcium hydroxide or $Ca(OH)_2$. This is also the chemical that is used to make the limewater solution that you use in the laboratory to test for carbon dioxide.

Experiment 6.2

Do bases affect dyes?

Erik remembered the experiment he did to find out about the effect that acids had on tea and beetroot water (page 71). He decided to find out what effect bases have on dyes.

This is what Erik recorded from his experiment.

Base	Effects on black tea	Effects on beetroot water
oven cleaner	darker	darker
washing powder	darker	darker
soap	darker	darker
shampoo	darker	darker

1 How do you think Erik did his experiment? Use these headings to show what you think he did:

Aim   What I need   Method   Conclusion

2 Now carry out your own experiment to find out whether three different bases have the same effect on dyes.

Activity 6.2

Identifying bases

1 What is the difference between a base and an alkali?

2 Give five examples of alkalis found in your home.

3 Choose five bases shown in the picture in Figure 6.5. Write them down in order from the strongest to the weakest.

4 Why should you not touch a strong base?

5 One difference between acids and bases is that acids taste sour while bases have a bitter taste. Why would it be dangerous to tell the difference between acids and bases by tasting them?

Unit 3 Using indicators

Indicators are special substances that turn one colour in an acid and another colour in a base. There are many different kinds of indicators. The most common indicators in school laboratories are litmus papers (blue and red) and phenolphthalein.

Litmus papers come in two different colours – red and blue. To test a substance, you use two pieces – one piece of red litmus and one piece of blue litmus:

- If the blue paper turns red, the substance is an acid.
- If the red paper turns blue, the substance is a base (remember the 'b' sound – blue is for base).
- If the papers do not change colour, the substance is neither an acid or a base – it is **neutral**.

Experiment

Classifying substances using litmus papers

You will need:
- small pieces of red and blue litmus paper ● test tubes and a rack
- a dropper ● samples of each substance listed in the table below

Substance	Red litmus paper	Blue litmus paper
lemon juice		
vinegar		
water		
caustic soda (drain cleaner)		
limewater		
milk of magnesia		
household cleaner with ammonia		

Method
Test each substance with both red and blue litmus paper. Copy the table and record any colour changes.

Questions
1 Why do you think you used water in this experiment?
2 How could you test solid substances (like soap) using litmus paper?
3 Does the litmus paper tell you how strong an acid or base is?

pH	Substance
0	battery acid, strong hydrofluoric acid
1	hydrochloric acid produced by the stomach lining
2	lemon juice, vinegar
3	grapefruit and orange juice, soft drinks
4	tomato juice
5	acid rain, black coffee
6	urine, milk, saliva
7	'pure' water
8	baking soda, sea water, eggs
9	toothpaste
10	milk of magnesia
11	household ammonia solution
12	soapy water
13	bleaches, oven cleaner
14	liquid drain cleaner, caustic soda

↑ Figure 6.6
The pH scale measures the strength of acids and bases on a scale from 1 to 14.

Strong and weak acids and bases

You have learned that acids and bases can be weak or strong. The strength of an acid or base depends on how it forms charged particles called **ions** when it is in solution. Acids form hydrogen ions and bases form hydroxide ions. The stronger an acid or base, the more ions it forms. Weak solutions of acids or bases produce very few ions.

Measuring the strength of acids and bases

The strength of acids and bases is measured on the **pH scale**. The pH scale has numbers from 1 to 14. Figure 6.6 shows you the pH scale and the strengths of some common acids and bases.

- Acids have a pH of less than 7. Strong acids have a low pH.
- Bases have a pH greater than 7. Strong bases have a high pH.
- Neutral substances have a pH of 7.

Universal indicator

In Figure 6.6, notice that each pH number matches a different colour. Universal indicator paper or solution is a mixture of indicators used to find the pH of substances. The paper or solution changes colour and the colour indicates the pH. A more accurate way to measure pH is with a special instrument called a pH meter.

Activity 6.3

Making sense of indicators and measurements

1 What is an indicator? Explain how you can use one to sort substances into three groups.

2 Why is universal indicator different from litmus paper indicators?

3 Write each substance shown in the box into the correct column of the table.

Strong acid	Weak acid	Neutral	Weak base	Strong base

blood (pH 7.4)   limewater (pH 10.5)   ammonia (pH 12)   battery acid (pH 1)   black coffee (pH 5)
oven cleaner (pH 14)   lemon juice (pH 2.3)   distilled water (pH 7)   toothpaste (pH 8)

4 Shampoo advertisements often say it is 'pH balanced'. What does this mean?

Neutralisation

You can make an acid neutral by adding a base to it. You can make a base neutral by adding an acid to it. When you mix acids and bases, they cancel out each other's effects. When this happens, we say that the acid or base has been neutralised.

During **neutralisation**, the acid and base combine to form a salt and water. This process can be written as a word equation like this:

acid + base → salt + water

Neutralisation can produce very useful substances from poisonous chemicals. For example:

hydrochloric acid + caustic soda → sodium chloride + water

Sodium chloride is a harmless substance. You probably know it better as table salt! Other useful salts that can be produced by neutralisation are ammonium phosphate which is used as a fertiliser, silver bromide used in photographic film and potassium nitrate which is used as a food preservative.

 **Applying your knowledge**

1 Write down the general equation for a neutralisation reaction.

2 Zuki has made a neutral solution by adding bicarbonate of soda (a base) to tartaric acid. Will her solution stay neutral if she adds more bicarbonate of soda? Give a reason.

3 In some rural areas people use ash from burned wood to clean their teeth. What does this tell you about the pH of ash?

4 Mrs Jenssen is growing hydrangeas. They all have blue flowers. She would like her plants to have both pink and blue flowers. What advice would you give her?

5 During an industrial accident a factory spilled a large volume of acidic liquid into a nearby lake.
   a) What effect would this have on the water in the lake?
   b) How would this affect the plants and animals in the lake?
   c) What could the company do to neutralise the effects of the acid in the lake?

Acid soils are not good for growing crops so farmers add lime to neutralise them.

Too much stomach acid can cause indigestion. We take antacid medicines to neutralise it.

Bee stings are acidic. They can be neutralised with soap or bicarbonate of soda.

Bluebottle stings are alkaline.

They can be neutralised by a weak acid such as vinegar.

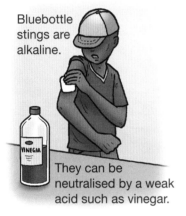

Acid from our mouth can cause tooth decay.

Toothpaste is alkaline to neutralise conditions in the mouth.

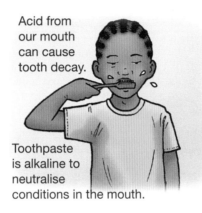

Ant bites are acidic. The pain can be relieved by calamine lotion.

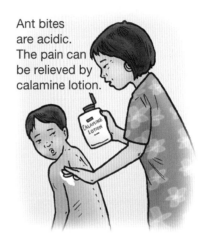

This fertiliser is a useful salt made by a neutralisation reaction.

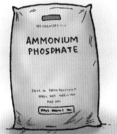

Hydrangeas produce pink flowers in alkaline soils, but blue flowers if you add acid.

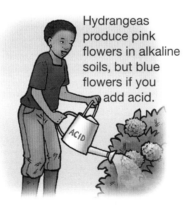

Adding a base like ammonia to an acid spill will neutralise it and prevent damage.

Some fire extinguishers mix an acid and a base to produce water, which is pushed out onto the fire.

In industry, bases are mixed with acidic gases before they are released into the air, so they do not cause acid rain.

↑ Figure 6.7 Some ways in which neutralisation is useful to humans

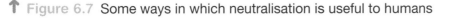

Chapter summary

Figure 6.8 provides a summary of the main concepts of this chapter.

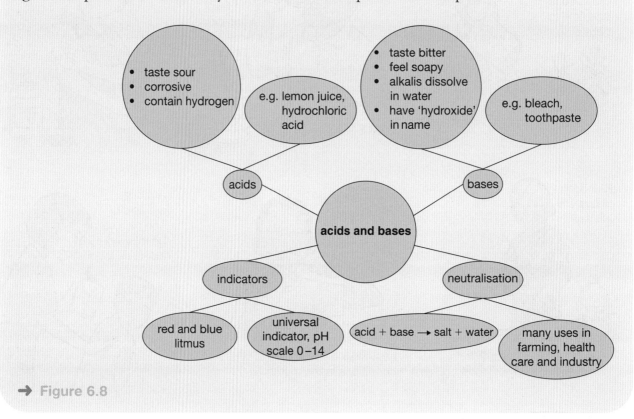

* taste sour
* corrosive
* contain hydrogen

e.g. lemon juice, hydrochloric acid

* taste bitter
* feel soapy
* alkalis dissolve in water
* have 'hydroxide' in name

e.g. bleach, toothpaste

acids

bases

acids and bases

indicators

neutralisation

red and blue litmus

universal indicator, pH scale 0–14

acid + base → salt + water

many uses in farming, health care and industry

➜ Figure 6.8

Revision questions

1 Give three examples of acids and three examples of bases found in everyday life.

2 Copy and complete these statements about the properties of acids and bases.
   a) Acids have a _____ taste.
   b) Acids turn _____ litmus paper _____.
   c) Acids have pH numbers _____.
   d) Acids react with bases forming a _____ and _____.
   e) Alkalis feel _____.
   f) Alkalis turn _____ litmus paper _____.
   g) Alkali solutions have pH numbers _____.
   h) All alkalis react with acids to _____ them.

3 Explain using your own words what pH means and what it tells you about substances.

4 A patient in a clinic complains of burning urine. The nurse tells her to drink some bicarbonate of soda (a base) mixed with water. Explain how this may help.

Physical and chemical changes

↑ **Figure 7.1** Can you see that there are solids, liquids and gases in these photographs?

The non-living things around us can be grouped into **solids**, **liquids** and **gases**. In this chapter you are going to describe these three **states of matter** and understand how substances can change from one state to another. You will learn the difference between physical and chemical changes. Then you will investigate the effects of heat on different substances and learn what happens during evaporation, burning and melting.

As you work through this chapter you will:

- use characteristics to distinguish between solids, liquids and gases
- describe changes from one state to another
- investigate what happens when things are heated
- measure and record temperature changes
- investigate evaporation
- explain the difference between physical and chemical changes, and give examples of each type
- describe what happens when substances burn or melt, and compare the two processes.

Unit 1 Solids, liquids and gases

You already know that we can divide the things on Earth into living and non-living things. And you know that we can classify living things according to their characteristics. Now you are going to find out how to use characteristics to group non-living things into solids, liquids and gases.

Solids, liquids and gases are found all around us. The land we live on is solid rock or soil. The oceans, rivers and lakes on Earth are filled with liquid water. The air that we breathe is a gas.

↑ **Figure 7.2** These items are all solids.

Each of these three groups has its own special characteristics. We use these to decide whether something is a solid, liquid or gas.

Solids

Solids have a fixed shape. You can pick up a solid and move it, or place it in a box or bowl and it will keep its shape. Solids do not change their shape unless they are forced to do so.

Liquids

Liquids do not have a fixed shape. A liquid will change shape depending on the container that you pour it into. You can pour liquids to move them from one place to another, but you cannot pick them up unless they are in a container.

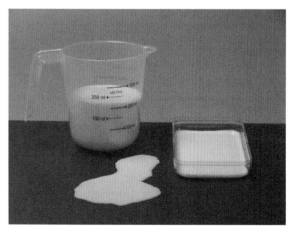

↑ **Figure 7.3** Milk is a liquid.

→ Figure 7.4
This liquid has changed its shape, but there is still the same amount. We say that liquids keep their volume.

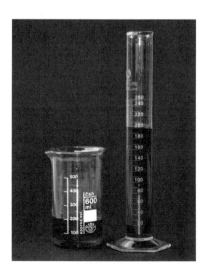

Gases

There are many different examples of gases. The air around us is made of many different gases (oxygen, nitrogen and carbon dioxide). Helium balloons are filled with gas. Some people use natural gas to cook.

Even though we are surrounded by gases, they are not as easy to see as liquids and solids. You can feel gases when the wind blows, you can smell them if they have a smell and you can hear them if they make a noise as they move, such as when air escapes from a car tyre.

Gases have no shape of their own. They have no fixed volume so they can spread out to fill the space they are in. They can also be squashed into a much smaller space.

↑ **Figure 7.5** Air (a gas) is squeezed into a bicycle tyre. When valve is released, the air spreads out into the space around the tyre and the tyre goes flat.

All non-living things have to be a solid, a liquid or a gas. Solids, liquids and gases are called the three **states of matter**. Figure 7.6 summarises the characteristics of each state of matter.

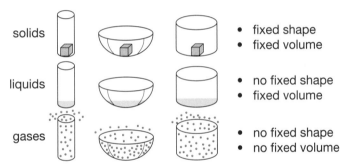

↑ **Figure 7.6** The characteristics of the three states of matter

Substances can change from one state to another. For example, solid water, or ice, can melt and change into a liquid, water. If the water is boiled, it changes into a gas – steam or water vapour. You will learn more about these changes of state in Unit 2.

 **Activity 7.1**   **Classifying objects as solids, liquids or gases**

1 Copy this table into your notebook. Write down the names of ten solids, liquids and gases you have come across in your everyday life.

Solids	Liquids	Gases

2 Which do you use most – solids, liquids or gases? Why?

Unit 2 Changing state

Non-living things can exist in three states – as solids, liquids or gases. But, if something is a solid, liquid or gas, it does not have to stay in that state. Substances can change from one state to another when they are heated or cooled. For example, water can change from a liquid to solid ice when it is cooled down in a freezer. When ice is taken out of the freezer and allowed to warm up, it changes from a solid back to a liquid. Water can also change from a liquid state to a gas state when it is heated and it changes to steam. Changes of state are temporary changes because they can be reversed by heating or cooling.

Melting

One way that substances change state is by melting. When heat changes a solid to a liquid we say it has melted. Every substance has a temperature at which it melts. This is called its **melting point**. Ice has a melting point of 0 °C. So, ice will melt and change state back to water at any temperature above 0 °C.

You can find the melting point of a substance by heating it and recording the temperature at which it changes from a solid to a liquid.

Experiment 7.1

Finding the melting points of substances

Aim
To measure and record the melting points of different substances.

You will need:
- a burner and a tripod ● a beaker and warm water ● a thermometer
- test tubes and tongs ● butter, sugar, chocolate and candle wax

Method
Place a beaker of water over a low flame to heat it slowly. Support a thermometer so that its bulb is under the water, but not touching the bottom of the beaker.

Place a sample of each substance in a separate test tube. Use the tongs to lower the test tube into the hot water. Do not let the test tube touch the bottom of the beaker.

When the substance starts to melt, check and record the temperature on the thermometer.

Repeat the steps for each substance.

Questions
1 How could you make your results more accurate?
2 Which substance has the highest melting point?
3 Which substance has the lowest melting point?
4 Why can you not use substances like metals and glass in this experiment?

Freezing

When a liquid changes to a solid, we say it freezes. The temperature at which a liquid freezes is called its **freezing point**. The freezing point of water is 0 °C.

→ **Figure 7.7** As you go higher, it gets colder. This is why there is snow on the top of these mountains, but no snow lower down. If the temperature gets above 0 °C at the top of the mountain, the snow will melt.

The melting point and freezing point of a substance are exactly the same temperature. But whether the substance melts or freezes depends on whether you are heating it or cooling it. Heating causes melting. Cooling causes freezing.

Boiling

Boiling some water causes it to change from a liquid to a gas. The temperature at which this happens is called the **boiling point**. For water, the boiling point is 100 °C.

If you heat water to 100 °C, the water changes from a liquid to a gas (steam). But, its temperature does not get any higher, even if you keep heating it. Once a liquid has reached its boiling point, the temperature stops increasing and remains constant.

Evaporation and condensation

You do not have to boil water to make it change state. If you put water in a dish outside in the sunshine, it will slowly disappear. This is because the water **evaporates** at temperatures lower than its boiling point.

If water vapour (gas) is cooled, it changes back into a liquid. This is called **condensation**. You can observe this happening if you take a cold drink from a refrigerator and leave it in the air. Water droplets form on the surface of the can or bottle because the water vapour in the air around the drink gets cold and it condenses.

Experiment 7.2

Comparing rates of evaporation

Aim
To find out whether heat affects the rate at which water evaporates.

Hypothesis
A wet handkerchief placed in a warm place will dry faster than a wet handkerchief placed in a cool place.

You will need:
- two wet cotton handkerchiefs or cloths

Method
Wring out the cotton cloths so that they are not dripping water.

Hang one cloth in a warm, sunny place.

Hang the other cloth in a cool, shady place.

Check the cloths every 5 minutes to see how they are drying.

Questions
1 Which cloth dried fastest?
2 Does this support your hypothesis?
3 What other factors could have affected the rate of evaporation?
4 How could you make sure these factors did not affect your experiment?

Sublimation

Some substances can change from a solid to a gas (and back again) without changing into a liquid state. This is called **sublimation**. For example, frozen carbon dioxide, or dry ice, changes directly from a frozen solid to a white smoky gas. You may have seen this on television during pop concerts. Sometimes the temperature in the air is so cold that water vapour freezes in the air and forms snow.

Charting results

The graph in Figure 7.8 shows how the temperature of some ice changed when it was warmed up in a tray in an oven.

Study the graph carefully. Read the labels and make sure you can see when:

- the ice reaches its melting point
- the ice changes state to a liquid
- the liquid reaches its boiling point
- the liquid changes state to a gas.

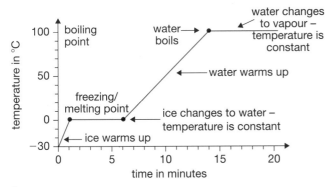

↑ **Figure 7.8** The temperature of ice as it is warmed in an oven

---

Activity 7.2   **Explaining changes of state**

1 How can a gas change to become a liquid and then a solid?

2 What does heating do to most solids?

3 What is the freezing point of water?

4 Why does food dry out if you leave it in a hot oven?

5 In a laboratory, a clear liquid boiled at 80 °C. How do you know this liquid is not water?

6 Match the letters A to F in Figure 7.9 with the correct term from the box.

freezing
dissolving
boiling
evaporating
melting
condensing

→ **Figure 7.9**

Unit 3 Physical and chemical changes

Physical changes

When you put candle wax in your test tube and heated it, in Experiment 7.1 (page 82), the wax melted. The wax changed state from solid wax to liquid wax. When the liquid wax cooled down again, it turned back into solid wax. This type of change is called a **physical change**.

In a physical change, the substance looks different, but it is still the same substance. The change is normally reversible and the substance can easily be changed back into its previous state. In a physical change, there is no change in mass. If you melt 10 grams of wax, you will get 10 grams of liquid wax. If you freeze 20 grams of water, you will get 20 grams of ice.

Changes of state are all physical changes. Other examples of physical changes are:

- breaking up a solid to make a powder
- mixing substances such as iron and sulphur
- dissolving substances such as sugar or salt in water.

Chemical changes

Figure 7.10 shows you what happens when you light a candle and let it burn.

➜ **Figure 7.10**
If you burn a candle, the wax and the wick disappear and you cannot get them back.

When you burn a candle, the wax seems to disappear. What actually happens is that the wax is broken down to form simpler substances. You cannot see these because they get carried away by the air. But if you hold a burning candle up against a beaker of cold water, you will see that black carbon forms on the glass. You will also see bubbles of water condensing on the glass. This type of change is called a **chemical change**.

You can usually tell that a chemical change has happened if a new substance has formed, and if the change cannot be reversed.

Some chemical changes can be reversed by chemical reactions, but most chemical changes are not reversible. For example, if you burn a match you cannot get the wood back.

In a chemical change, the structure and make-up of the original substance is changed and new materials are made. Because the substances change, the mass of the substances the also changes. Chemical changes often give out large amounts of energy in the form of heat, light and electricity. A burning candle gives out heat and light. A battery in a torch gives out electricity as the chemicals inside the battery mix and change.

Other examples of chemical changes are:

- toasting bread or frying an egg
- carbon dioxide mixing with limewater to make it milky
- rusting iron
- heating iron and sulphur together to make a new substance called iron sulphide.

Activity 7.3   **Identifying physical and chemical changes**

Say whether each picture in Figure 7.11 shows a physical or a chemical change. Give a reason for each answer.

↑ **Figure 7.11**

Chapter summary

✓ Non-living things can exist as solids, liquids or gases. These are the three states of matter.

✓ Solids, liquids and gases have characteristics that make them different from each other.

✓ Substances can change from one state to another when they are heated or cooled.

✓ The processes involved in changes of state are melting, freezing, boiling, evaporation, condensation and sublimation.

✓ Changes of state are physical changes.

✓ A physical change is a reversible change. The substance looks different but it is still the same substance.

✓ A chemical change is not easily reversible. In a chemical change new substances are formed and energy is used up or given out. Burning is a chemical change.

Revision questions

1 Draw labelled diagrams to show the differences between solids, liquids and gases.

2 What happens to water when you boil it?

3 What happens to ice when you melt it?

4 Normal room temperature is about 25 °C. Say whether the following substances would be solids, liquids or gases at room temperature.

Substance	Melting point (°C)	Boiling point (°C)
water	0	100
tin	230	2260
butane	−135	−1
benzoic acid	125	250
oxygen	−220	−183

5 Complete these sentences by filling in the missing words:
   a) When chocolate melts in the sunshine, it is a _____ change. The melted chocolate turns back into a _____ when it cools.
   b) Burning wood is a _____ change. New substances are formed and you cannot get the _____ back.

Materials and their properties

↑ **Figure 8.1** Some of the materials we use every day

Different things are made from different kinds of materials. These materials are often chosen because of their **properties**. Some materials need to be light, some strong, some waterproof, some heat-resistant and some need to be see-through (transparent).

In this chapter, you are going to learn about the different properties of materials that make them useful to us. You will also learn to group materials as metals and non-metals.

To understand and investigate materials and their properties, you need be able to:

- identify materials and give examples of their properties
- compare waterproof and absorbent materials
- describe materials that are brittle, flexible or malleable
- tell the difference between opaque and transparent materials
- classify materials as metals and non-metals.

Unit 1 Using materials

When we make things, we have to decide which materials to use. Different materials behave in different ways. For example, metals can bend, some plastics melt when they are heated, and certain fabrics are waterproof. We can say that different materials have different **properties**. You have to think about which properties are important when you choose a material. It would be silly to use materials that are not waterproof to make an umbrella or to use materials that melt easily when you are making a heater.

Choosing materials for different uses

Look at the different materials used to make things in Figure 8.2.

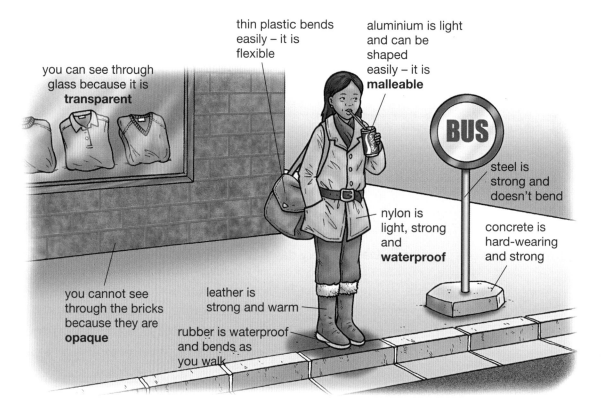

you can see through glass because it is **transparent**

thin plastic bends easily – it is flexible

aluminium is light and can be shaped easily – it is **malleable**

steel is strong and doesn't bend

nylon is light, strong and **waterproof**

concrete is hard-wearing and strong

you cannot see through the bricks because they are **opaque**

leather is strong and warm

rubber is waterproof and bends as you walk

↑ **Figure 8.2** Can you say why these materials have been used to make these objects?

The materials in the picture have been chosen because they have certain properties.

Materials that are found in nature are called **natural materials**. Those that are made by humans are called manufactured or **synthetic materials**.

Properties of different materials

Figure 8.3 summarises some of the properties of everyday materials.

Material	Properties
Plastics	 Some plastics melt when they are heated, others become hard and brittle Some are flexible and easily bent Strong and light Can be moulded and shaped easily Waterproof Some plastics are transparent, some are opaque
Textiles	 Most are absorbent (soak up water) but some are waterproof Strong fibres give them strength
Wood	 Hard and strong Can float on water (low density) Opaque Burns when heated Plywood is very strong and can be bent (flexible)
Glass	 Hard and brittle (breaks easily) Transparent (but can be coloured, making it translucent) Waterproof
Ceramics	 Hard and brittle Able to withstand heat Waterproof Opaque
Metals	 Shiny, hard and strong Malleable (can be shaped easily) Can be made into wire (ductile) Melt at high temperatures Some are magnetic

➜ Figure 8.3 Properties of materials

Activity 8.1   **Choosing materials for different jobs**

1 Write down the material that you would use to make each item in the box below. Figure 8.3 might help you decide.

2 What properties are important for making each item?

floor boards   food cans   electric cable   table tennis bats   desks   shelves
furniture   nails   trays   cool-drink bottles   bleach bottles   jerseys   raincoats
socks   jeans   carpets   watch faces   syringes   buckets   machine parts

Unit 2 Investigating the properties of materials

Absorbent materials

→ **Figure 8.4**
These materials all absorb water.

Absorbent materials are able to take in liquids easily. All absorbent materials have tiny holes or pores in them which allow liquids to pass through. They also have spaces inside them that allow them to collect and store liquids. You can see these holes clearly in the sponge in Figure 8.4, but you cannot see them so easily in the other materials.

Paper towels, tissues, toilet paper, cotton wool and blotting paper are all absorbent. Many textiles are absorbent – that is why your clothing gets soaking wet if you stand in the rain.

Bricks and some soft rocks are absorbent, so they can hold water. In a brick house, this is not a good property. The water from the ground moves up through the bricks and makes the walls damp. Builders can stop this by making air spaces in the foundations of the house and by putting a layer of waterproof, rubbery material above the bricks in the foundations to stop water rising and damp spreading up the walls.

Waterproof materials

Waterproof materials stop liquids from passing through them at all. Glass, metal and plastic are waterproof.

Some materials, like the material used to make rubber gloves, wetsuits and fire-fighters' jackets are **water-repellent**. These materials do not have holes or pores, so water cannot pass through them at all. Even the holes made by needles when the material is sewn together are sealed with a special coating or by putting a layer of waterproof tape on the inside. When these materials come into contact with water, the water just runs off them.

Other materials, like the nylon used to make tents, are **water-resistant**. These materials are normally made from woven threads.

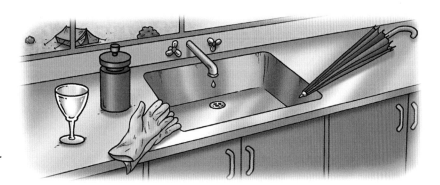

→ **Figure 8.5**
These materials are waterproof, water-repellent or water-resistant.

There are small spaces between the threads, and water could get through the material. To stop water from going through the small spaces, the material is treated with special substances, such as silicone, to make the water run off. These materials will get soaked through if there is enough water.

Transparent and opaque materials

Transparent materials are see-through. They allow light to pass through them without any interference. Glass, clear plastic, air and water are all transparent.

Opaque materials are not see-through. They do not allow light to pass through them. Wood, metal, stone and most fabrics are opaque. Some fabrics are quite thin, and they allow some light to pass through them, so you can see through them if you look closely. These materials are called **translucent** materials. Coloured glass and glass with patterns on it are translucent. These translucent types of glass are often used in bathroom windows to make the rooms more private.

Activity 8.2 **Classifying materials found at school**

1 Make a list of 20 different materials used to make things at your school.

2 Copy the table to show the properties of each material on your list. Tick the columns to show which properties each material has. Make sure you leave enough space for all the items on your list.

Material	Absorbent	Waterproof	Transparent	Opaque	Translucent

3 What does it mean if an advert says a watch is water-resistant to 30 metres?

Unit 3 Bending, shaping and breaking materials

When we use materials to build roads or bridges, we need to know how strong they are, and how they will behave if we bend them, push them, pull them or hit them.

There are four properties of materials that are important when you are thinking about strength.

Flexible or stiff?

A **flexible** material is one that bends easily when you apply a force to it. A **stiff** material does not bend easily.

The flexibility or stiffness of a material can be linked to its shape. A thin sheet of metal may be flexible, but a thick sheet may be stiff.

↑ **Figure 8.6** These materials are all flexible.

↑ **Figure 8.7** These materials are not flexible, they are stiff.

Malleable and ductile

Malleable means easy to press or bend into a new shape. You can shape malleable materials without cracking or breaking them. Putty, clay and baking dough are malleable – you can press them, roll them and mould them. Hot metals are malleable. When jewellers make objects with gold and silver, they heat the metal up before they bend or shape it, to make it more malleable.

➜ Figure 8.8 Metals are ductile, so they can be used to make wire.

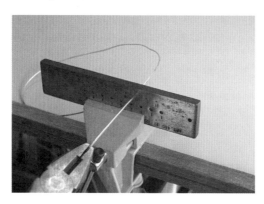

Ductile means able to be pressed or pulled into a shape without breaking or cracking. Ductile metals, such as silver, can be pulled through a small hole and drawn out into thin wire without cracking or breaking (Figure 8.8)

Brittleness

Brittle materials crack or break easily if you bend them or hit them. Glass is very brittle – so are eggshells, pottery and cast iron. Rocks and plain concrete are also brittle – if you hit them hard with a hammer, they shatter.

Experiment 8.1

Which sweets are most brittle?

Design and carry out an investigation to compare the brittleness of five different kinds of hard sweets or biscuits.

Write down what you do and what you find out.

Activity 8.3 Linking properties to uses

1 Study the picture of the car in Figure 8.9.

→ Figure 8.9

Give examples of materials used to make the car that are:
a) transparent b) opaque c) stiff d) flexible
e) malleable f) ductile g) brittle.

2 Find out what is done to the glass used in car windows to make it less brittle. Tell the class what you find out.

Unit 4 Metals and non-metals

Materials can be classified as metals or non-metals depending on their properties.

Metals

Metals are very common in everyday life. Iron nails, wire fences, aluminium foil, jewellery, cooking pots and tools are all made from metals.

➡ **Figure 8.10**
These objects are all made from metals.

Metals are useful because of their properties:

- Most metals are solid at room temperature.
- Metals are shiny when they are polished. When the surface of a metal becomes dusty or oily it does not look shiny, but once it is cleaned, it has a shine.
- Metals are malleable.
- Metals are ductile.
- Metals are good conductors of heat (see Chapter 12) and electricity (see Chapter 13).

The table on page 97 shows you some of the properties and uses of metals that you know.

Non-metals

Non-metals do not have any of the properties of metals. They are generally poor conductors of heat and electricity. Solid non-metals are not malleable or ductile and they are usually very brittle.

Metal	Properties	Uses
aluminium	soft, light, resistant to corrosion, does not rust	drink cans, tin foil, power lines, boat hulls, aircraft panels
copper	malleable, soft, good conductor of heat and electricity, resistant to corrosion, does not rust	water pipes and tanks, electrical wiring, jewellery and ornaments
gold	the most malleable and ductile metal, soft, does not react with air or other substances	jewellery, electronic components, dental fillings
iron	hard, magnetic, high melting point	used to make different kinds of steel, cast iron
lead	soft, malleable, very low melting point	car batteries, lead sheeting used in building, used to make solder, fishing weights and sinkers
nickel	hard, magnetic, corrosion resistant	used to coat coins and electroplate jewellery and other metals, added to glass to give it the green colour
silver	malleable and ductile, soft, low melting point, good conductor of electricity, does not rust	jewellery, cutlery, coins, solders and brazing alloys, electrical contacts
tin	malleable and ductile, soft, corrosion resistant	tin plates, coating for other metals
zinc	hard, low melting point, resistant to corrosion, does not rust	dry batteries, roof cladding, galvanising, light coins

Activity 8.4   Identifying metals

1 Name five metal items that you use every day.

2 Why do you think each of these items was made from metal rather than another material?

3 Find out the names of the metals used to make each item.
   a) Use the table to find the properties of each metal used.
   b) Explain how these properties make each metal well suited to its use.

4 While you are out on a walk, you find a strange material. You clean it and it is not shiny. Then you hit it with a hammer and it breaks into small pieces. Do you think this is a metal? Give a reason for your answer.

Chapter summary

- ✓ Wood, metals, glass, ceramics, fabrics and plastics are all materials.
- ✓ Different materials have different properties that make them useful for different purposes.
- ✓ Absorbent materials like towelling and paper take in water.
- ✓ Waterproof materials like glass and rubber do not take in water.
- ✓ Transparent materials are see-through because they allow light to pass through them.
- ✓ Opaque materials are not see-through because they do not allow light to pass through them.
- ✓ Translucent materials allow some light to pass through them.
- ✓ Flexible materials bend easily. Stiff materials do not bend easily.
- ✓ Malleable materials can be shaped easily.
- ✓ Ductile materials can be pulled and shaped into wire.
- ✓ Brittle materials break easily if they are bent or hit.
- ✓ Most materials can be classified as either metals or non-metals.
- ✓ Metals are shiny, malleable and ductile and they conduct heat and electricity. Non-metals do not have these properties.

Revision questions

Explain in your own words what these words mean and give an example of each.

1 natural material

2 synthetic material

3 absorbent

4 waterproof

5 water-repellent

6 water-resistant

7 transparent

8 opaque

9 translucent

10 flexible

11 stiff

12 malleable

13 ductile

14 brittle

15 metal

16 conductor of heat

17 conductor of electricity

18 non-metal

↑ **Figure 9.1** Ice, water and steam are all the same substance, so why do they look and behave so differently?

You know that solids, liquids and gases have different properties, but do you know why?

Scientists use **models** to help explain how things happen in science. One of these models, called the particle model, helps to explain why solids, liquids and gases behave in different ways.

In this chapter, you are going to learn about the particle model. You will then investigate solids, liquids and gases to find out what happens to the particles when they are heated or cooled. You will learn more about the properties of ice and water, and explain what happens when water freezes. You will then investigate gases in the air to understand more about pressure and how gases spread through air.

As you work through this chapter, you will:

- use the particle model to describe solids, liquids and gases
- investigate expansion in solids, liquids and gases
- explain and describe what happens when water freezes and observe the effect of salt on the freezing point
- use the particle model to explain dissolving
- show how gases in the air cause pressure
- explain how gases spread through the air by diffusion.

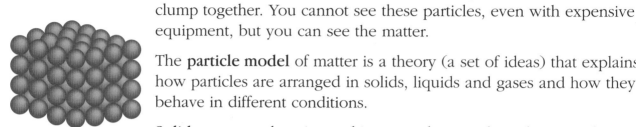

Unit 1 The particle model

All solids, liquids and gases around us are made of **matter**. Scientists believe that matter is made of tiny particles (pieces or bits) that clump together. You cannot see these particles, even with expensive equipment, but you can see the matter.

The **particle model** of matter is a theory (a set of ideas) that explains how particles are arranged in solids, liquids and gases and how they behave in different conditions.

Solid matter, such as ice and iron, can keep a shape because the particles in the solid are packed very closely together. This stops them from moving around from place to place. The particles in a solid are held together by very strong forces.

Liquid matter, such as water and petrol, cannot keep a shape because the particles are held together by weaker forces so they can move around. This is why liquids take the shape of the container they are poured into – the particles move around to fit the new shape.

In **gases**, such as steam and air, the particles are spaced out and far apart. The particles can move freely in any direction. The particles bump against each other as they move. When a gas is placed in a closed container, the particles keep moving and they bump against the sides of the container, but they cannot escape unless the container is opened.

Figure 9.2 shows how particles are arranged in solids, liquids and gases. The table summarises the particle model of solids, liquids and gases.

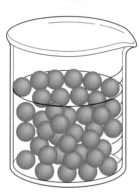

solid

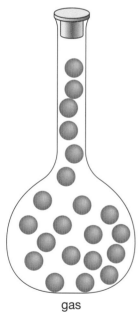

liquid

gas

↑ **Figure 9.2**
How particles are arranged in solids, liquids and gases

State	Spacing of particles	Movement of particles	Forces holding particles together	Shape of matter
solid	tightly packed, very close together	cannot move around	very strong	fixed shape
liquid	close together, but loosely packed	able to move past each other	weaker than in solids	no fixed shape, takes shape of a container
gas	spread out, far apart	free to move in any direction	very weak	no fixed shape, fills space in a container

↑ **Figure 9.3** Looking at sand can help us to understand the particle model.

Thinking about sand may help you to understand the particle model. Look at the photographs in Figure 9.3.

In a solid block of sandstone, the grains of sand (think of them as particles) are tightly packed together and they cannot be separated. You cannot squash the sandstone by pushing it, and it keeps its shape.

Loose sand behaves more like a liquid. The sand still contains grains (particles) that are close together, but they are not strongly joined to each other. When you pour the sand, the grains move around and slide over each other, so the sand can be poured from a bucket and it will take the shape of a container.

When the wind blows, the sand grains get spread out and carried around far from each other. They are able to move around like the particles in a gas.

 **Activity 9.1**   **Explaining the particle model**

1 Draw diagrams to show how particles would be arranged in ice, water and steam.

2 How would you answer these questions, using the particle model?
   a) Why can you squash a gas but not a liquid?
   b) Why can you pour liquids but not solids?
   c) Why do solids keep their shape?
   d) What happens when you break a solid?
   e) Why does a gas fill a container completely?

Unit 2 Expansion and contraction

Expansion

When solids, liquids or gases are heated, they take in heat energy. The heat energy makes the particles **vibrate** (move up and down) more. When the particles start to move like this, they spread out and they take up more space so the substance gets bigger. We call this getting bigger **expansion**. For example, when you take a deep breath, your ribs expand. When a substance expands, its volume increases.

You can do a simple experiment to help you understand what happens when solids, liquids or gases are heated.

Experiment 9.1

Observing particle movements

Aim

To demonstrate how energy makes particles move and spread out.

You will need:
- a glass jar with a lid
- a handful of dried beans or lentils

Method

↑ **Figure 9.4** What happens when you shake the jar?

Shake the jar very gently. The beans move around a bit. This is what particles of matter normally do in a solid.

Shake the jar a bit harder. Notice that the beans separate and take up more space in the jar. This is what happens when matter is heated and the solid melts to form a liquid.

Now shake the jar really hard. The beans seem to want to move out of the jar. They are much further from each other and they take up the whole space in the jar. This is what happens when liquids reach their boiling point. The particles with the most energy escape from the surface of the liquid into the air as a gas.

Contraction

Contraction is the opposite of expansion. It means getting smaller or tighter. When a substance contracts, its volume decreases.

When you cool a liquid, the particles lose heat energy and they cannot move as freely. When the liquid loses so much energy that the particles cannot move around any more, it freezes and becomes a solid.

When water freezes (at 0 °C) the particles in the water begin to clump together like the particles in a solid. Ice has fewer particles for the same volume than liquid water, so ice is lighter than water. That is why ice blocks float in a glass of water.

Water has one property that makes it slightly different from other liquids. When water cools, it contracts like other liquids till the temperature is 4 °C. From temperatures of 4 °C to 0 °C, the water actually expands. The particles spread out and the water becomes less dense.

This means that cold water (from 4 °C to 0 °C) will rise to the surface. When the temperature gets to 0 °C, the water freezes and turns to ice. Because the colder water is at the surface of lakes, reservoirs and ponds, ice usually forms here first. The water below may not freeze at all, so fish and plants can survive in very cold places because the water they live in does not freeze.

The water in the oceans does not usually freeze. This is because salt water has a lower freezing point than fresh water. Figure 9.5 shows you how this happens.

→ **Figure 9.5**
The temperature of the ocean is below the freezing point of pure water, but the salt water is above its freezing point, so it stays in a liquid state.

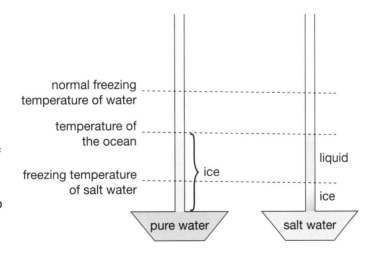

Do liquids expand as much as gases?

Aim

To see how a liquid expands when it is heated.

You will need:
- a conical flask with a tube and a stopper ● a bowl of hot water

Method

Fill the flask to the top with water before you fit the stopper and the tube. Make sure some water rises up the tube. Mark the level with a felt-tip pen or a piece of tape.

Seal round the stopper and tube with petroleum jelly to make an airtight seal.

Stand the flask of water in the bowl of hot water, as shown in Figure 9.6. Observe what happens to the level of the water in the tube.

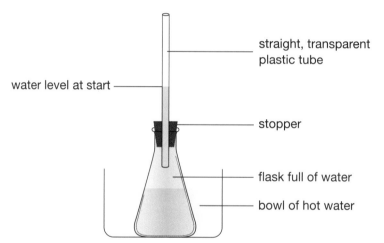

straight, transparent plastic tube

water level at start

stopper

flask full of water

bowl of hot water

⬆ **Figure 9.6** How to set up your equipment looking at the expansion and contraction of water.

Questions

1 What happens to the water level in the tube? Why?
2 Do you think liquids expand as much as gases when they are heated? Give a reason for your answer.

The particle model is useful to explain what happens in this experiment. The particles in a liquid are closer together than the particles in a gas. They also move more slowly. When you heat the water, the particles take in heat energy and they move faster. This makes the liquid expand and move up the tube, but there is far less expansion and contraction in liquids than in gases.

Expansion in solids

There are many different experiments that you can do to observe how solids expand when they are heated and contract when they are cooled. Figure 9.7 shows you two different pieces of equipment that can be used.

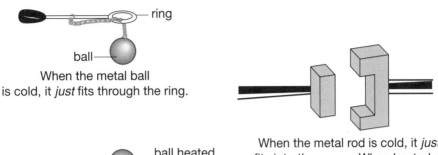

When the metal ball is cold, it *just* fits through the ring.

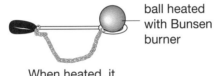

ball heated with Bunsen burner

When heated, it expands – and no longer fits.

When the metal rod is cold, it *just* fits into the gauge. When heated, it expands – and no longer fits.

➡ **Figure 9.7**
Equipment to show expansion and contraction in solids.

Because solids expand when they are heated, scientists and engineers have to think about what will happen to the materials they use in buildings and machinery if they get hot. They build expansion spaces or joints into roads, bridges and railway lines so that the structures will not crack or break if they expand and contract.

Activity 9.2 **Explaining expansion and contraction**

1 Why do solids, liquids and gases expand when they are heated?

2 Why do solids, liquids and gases contract when they are cooled?

3 Why does the water in a pond freeze from the top down?

4 Will salt water turn to ice at 0 °C? Give a reason for your answer.

5 Why do electrical power lines hang down more in summer than in winter?

6 Why do you sometimes find rubber strips between the sections of a concrete bridge?

Unit 3 Pressure

Solids, liquids and gases can press against surfaces. If you pick up a brick, it presses down against your hand. If you hold your hand under running water, the water presses against your hand. If you blow onto your hand, you can feel the air pressing against it.

In science we use the word **pressure** to talk about how particles push or press against surfaces. Pressure measures how strong a push is on a particular area of surface.

The pressure in a liquid increases towards the bottom of a container (or as you go deeper in the ocean). This is because more water is pressing down on the lower layers. Look at Figure 9.8.

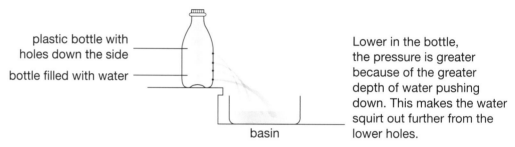

plastic bottle with holes down the side

bottle filled with water

Lower in the bottle, the pressure is greater because of the greater depth of water pushing down. This makes the water squirt out further from the lower holes.

basin

↑ **Figure 9.8** Pressures at the bottom of a liquid are higher than pressures at the surface.

Compression

When you squeeze something you are putting pressure on it. If the pressure makes the substance squash together, we say it has been **compressed**.

Solids cannot be compressed because the particles in solids do not have spaces between them. If you push down on your desk with your hand, it will not get a dent in it.

There are some spaces between the particles in liquids, but it is very difficult to compress them. If you cool a liquid, the particles slow down and move closer together and you can compress it slightly. Remember, though, that this doesn't work with water, because water expands as it freezes.

Gases are easy to compress. The particles in a gas are spread out, so you can force them to move closer together and compress the gas.

When you push air into a container, such as a bicycle tyre, you are pushing in more air particles. The particles keep moving and they bump against the walls of the tyre. As you push more air into the tyre, there are more particles to press against the sides of the tyre, and the air pressure in the tyre increases. The tyre gets hard because the air pressure inside the tyre is greater than the air pressure outside.

Atmospheric pressure

The **atmosphere** is the name given to the gas that surrounds the Earth. This gas presses down on the Earth's surface and exerts pressure on it. The pressure at the Earth's surface is higher than the pressure up in the sky because there is more air pressing down on the surface. The pressure at sea level is higher than the pressure on top of high mountains for the same reason.

The atmosphere around us presses on our bodies all the time. So, why don't we feel it? We don't feel air pressure because there is air inside our bodies and outside our bodies. The air inside our bodies pushes outwards and balances the pressure of the air around us which pushes inwards.

Pressure affects liquids and gases

When gases are put under very high pressures, the particles are forced together and the gas may change into a liquid state. Oxygen in tanks used for welding is normally a liquid under very high pressure. When the valve on the gas tank is opened, some of the pressure is released, particles escape from the liquid and oxygen comes out of the tank as a gas.

Pressure also affects the boiling point of liquids. When the air pressure on the surface of a liquid is lower, it needs less energy to make particles escape and evaporate, so the liquid boils at a lower temperature. This is why water boils at temperatures below 100 °C at the top of high mountains.

 **Activity 9.3** Applying what you have learned

1 What would happen to a blown-up balloon if you took it up a high mountain? Give a scientific explanation for this.
2 When you put your finger over the nozzle of a syringe and try to push the plunger in, it is difficult to do this. Explain why.
3 Why are motorists told to check the pressure in car tyres while the tyres are cold – in other words, before they go on a long journey?

Unit 4 Dissolving and diffusion

Dissolving

↑ **Figure 9.9** Coffee powder dissolves in hot water.

↑ **Figure 9.10** Nail polish dissolves in nail polish remover.

Dissolving normally involves mixing a solid and a liquid. When the solid dissolves in the liquid, the particles of the solid are pulled apart. These particles move into the small spaces between the particles of the liquid. The mixture of a liquid and a dissolved solid is called a **solution**.

Dissolving is not the same as melting. Melting is a change of state in which one substance changes from a solid to a liquid on its own. Dissolving always involves *two* substances. Coffee powder does not melt in water. It mixes with the water and it can be separated from the water again. You will learn more about this in Chapter 10.

Experiment 9.3

Investigating dissolving

Aim
To demonstrate how substances dissolve in water.

You will need:
● 1 teaspoon of coffee powder ● a beaker ● warm water

Method
Half fill the beaker with warm water. Stir the teaspoon of coffee powder into the water. Observe what happens.

Questions
1 What happened to the colour of the water? Why?
2 Draw a diagram to show what you think happened to the particles of coffee.

Diffusion

Imagine you are sitting at the back of the classroom and that a pupil at the front of the class sprays a small amount of perfume into the air. After a short time, you can smell the perfume in the air. How does this happen?

→ **Figure 9.11**
Perfume smells can spread through the air.

Perfume particles spread through the air by **diffusion**. Diffusion happens when particles of a substance spread out from an area where there are lots of the particles to an area where there are fewer of the particles. Particles will carry on diffusing until they are spread out evenly.

When two gases (such as perfume and air) or two liquids (such as water and food dye) are mixed in a container, the particles of the two substances will mix with each other until they are evenly spread through the container.

Activity 9.4   Explaining dissolving and diffusion

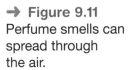

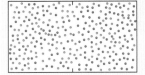

↑ **Figure 9.12**

1 Your younger sister stirs a teaspoon of sugar into a cup of tea. She believes that the sugar has disappeared. Explain to her why this is not really true.

2 When you get home after school, you open the door and you can smell burned food from the kitchen. Explain how the smell has spread from the kitchen to your nose.

3 The first diagram in Figure 9.12 shows two containers of different gases separated by a glass slide. In the second and third diagrams, the glass slide has been removed.

a) What is happening in the second and third diagrams?
b) What would you be able to observe in this experiment if one gas was coloured yellow and the other was colourless?

Chapter summary

✓ The particle theory of matter is a model which tries to explain the behaviour and properties of solids, liquids and gases.

✓ Solids, liquids and gases have different properties which can be linked to the way that their particles are arranged.

✓ Most solids, liquids and gases expand when they are heated and contract when they are cooled.

✓ Pressure is caused by particles in matter pressing down or pressing against the sides of the container they are in.

✓ Pressure increases with depth and is higher at sea level than at high altitudes.

✓ Solids cannot be compressed, liquids can be slightly compressed if they are cooled, and gases can be compressed easily.

✓ When a solid dissolves in liquid, its particles are pulled apart and mix with the particles of the liquid. The solid changes state, but it is still there and it can be recovered from the liquid.

✓ Melting is not the same as dissolving. Melting is a change of state in which only one substance is involved.

✓ Diffusion is the spreading out of particles from an area of high concentration to areas of low concentration to reach an even spread.

Revision questions

1 Use the particle theory to explain why:
   a) balloons get round and look solid when they are blown up
   b) you can smell cooking through a whole house
   c) chocolate is hard, but it forms a puddle when it melts
   d) you cannot see sugar when you stir it into a cup of warm water.

2 The beakers in Figure 9.13 show what happens to a solid when it is heated. Copy the diagram and write labels for each beaker to explain what state the substance is in and what is happening to it. Use what you know about particle theory to explain what is happening.

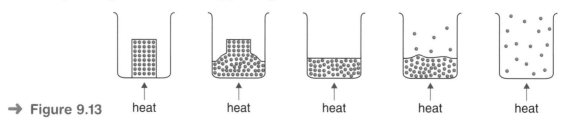

➜ Figure 9.13   heat   heat   heat   heat   heat

Mixing and separating substances

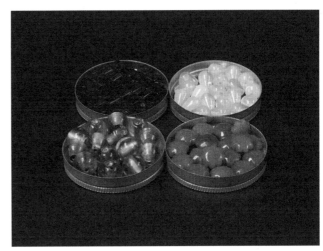

↑ **Figure 10.1** These beads are mixed together on the left but they can be separated and put into small groups, as on the right.

All the materials around us are either pure substances or mixtures. In this chapter, you are going to learn more about mixtures. You will learn what scientists mean when they talk about mixtures and you will find out about different kinds of mixtures. Then you will learn how to separate the substances in a mixture using magnets, filters, evaporation, distillation and chromatography.

When you have worked through this chapter, you will be able to:

- explain what a mixture is
- give examples of different types of mixtures, including solutions
- make a saturated solution
- explain how temperature can affect the solubility of different substances (how easily they dissolve)
- use different methods to separate mixtures
- discuss how separation of mixtures is used in everyday life.

Unit 1 Mixtures

Everything around us (matter) can be divided into two groups: pure substances and mixtures.

A **mixture** is a combination of liquids, solids or gases that can be broken down into the pure substances in the mixture by physical processes. For example, mud is a mixture of soil and water. We can separate the soil and water using filter paper to trap the soil. We can also separate the mixture by making the water evaporate, leaving the soil behind.

Pure substances cannot be separated or broken down into smaller parts by physical processes. Some pure substances, called **compounds**, *can* be broken down by chemical processes.

Types of mixtures

Mixtures can be any combination of solids, liquids and gases.

Gas-with-gas mixtures

Air is a mixture of gases. Gases such as oxygen, nitrogen and carbon dioxide can move around. When two or more gases meet each other, they mix. Another example of a gas-with-gas mixture is the mixture of anaesthetic gases that doctors use to make people unconscious during operations.

Gas-with-liquid mixtures

There are two ways in which gases and liquids can form mixtures.

- Tiny drops of liquid can be mixed with and carried in a gas. You can see an example of this when you use an aerosol spray (Figure 10.2). The 'spray' that comes out of the canister is tiny droplets of liquid carried in a gas.

- Gas droplets can be trapped inside a liquid where they form a foam. You can see this when you wash your hands with soap – bubbles of air get trapped inside a mixture of soap and water to make foam. Shaving foam and whipped cream are foams.

Gas-with-solid mixtures

↑ **Figure 10.2** Gas can mix with liquid to form an aerosol spray or a foam.

Small particles of solids can mix with gas, and be carried in the gas. The smoke from a coal or wood-burning fire is a mixture of gas fumes, pieces of burned material and ash.

Solid-with-solid mixtures

Solid pieces of one substance can mix with solid pieces of another substance. For example, you can make a mixture of sugar and salt, or herbs and spices. The soil is a mixture of rock particles and organic materials such as decaying leaves and dead insects.

Solid-with-liquid mixtures

If you stir talcum powder or fine sand into water you get a mixture. The particles of talcum powder or sand float in the water. We say they are suspended (hanging) in the water. This type of mixture is called a **suspension**.

If you stir sugar or salt into water you also get a mixture. But in this case, the solids dissolve in the liquid. You cannot see them any more, but if you taste the water, it will taste sweet or salty. This type of mixture is called a **solution**.

You will learn more about suspensions and solutions in Unit 2.

Liquid-with-liquid mixtures

Some liquids mix easily with other liquids. For example, ink will mix easily with water. Liquids which mix easily are called **miscible** liquids.

Other liquids do not mix so easily. For example, oil does not mix with water. When you pour oil into water, the oil floats on top of the water. Liquids which do not mix are called **immiscible** liquids.

When tiny drops of immiscible liquids are mixed evenly through another liquid we call the mixture an **emulsion**. Margarine is an emulsion of oil and water. Mayonnaise is an emulsion of oil and vinegar.

↑ **Figure 10.3** Oil and water are immiscible – they do not mix.

Activity 10.1 **Describing mixtures**

1 What does the word 'mixture' mean in science?

2 What type of mixture is each of these?

    a) iron filings and sand b) fizzy cola c) butter
    d) sea water e) soda water f) salt and pepper
    g) salad dressing h) cooking gas i) smoke from a factory chimney
    j) milk k) human tears l) shaving foam

Unit 2 More about solid-with-liquid mixtures

Solutions

You already know that sugar will 'disappear' if you stir it into a cup of water. You also know that the sugar is still there because the water tastes sweet.

The sugar and water have formed a mixture called a solution. The particles of sugar have dissolved and they are so small that you cannot see them, so the solution looks clear. If you shine a light on it, the light will pass through the solution.

Substances like sugar, which dissolve in water to form a solution, are described as **soluble** substances. Salt and sugar are both soluble in water.

Suspensions

When you stir chalk or fine sand into water they do not dissolve. We say they are **insoluble** and that they are in suspension. The particles in a suspension are normally big enough for you to see them. If you leave the suspension to settle, the solid will eventually gather at the bottom of the container.

⬆ **Figure 10.4** Both these beakers contain a mixture of clay and water. In the beaker on the right, it has been stirred to form a suspension.

**Experiment
10.1**

Is it soluble?

Aim

To test some substances to find out which are soluble and which are insoluble in water.

You will need:

- six test tubes and a test tube rack
- salt, sugar, sand, copper sulphate, copper carbonate, chalk
- water
- a teaspoon

Method

Half fill each test tube with water. Add a quarter teaspoon of salt to the first test tube. Shake the test tube gently.

Examine the test tube as you shake it. Notice whether the liquid is clear or cloudy (whether light passes through it or not), and whether it is coloured or colourless.

Place the test tube in the rack and allow it to stand for five minutes.

Repeat the steps for each of the other substances. Record your observations in a table like this one.

Mixture	What happens at first?	What happens after five minutes?	Solution or suspension?
salt and water			
sugar and water			
sand and water			
copper sulphate and water			
copper carbonate and water			
chalk and water			

Questions

1 Did all the substances dissolve?
2 What do you think would have happened if you had added five teaspoons of each substance? Why?

Solvents and solutes

The liquid in a solution is called the **solvent**. The solid substance which is dissolved in the solvent is called the **solute**.

Water is the most common solvent. More substances will dissolve in water than in any other solvent. But not all substances will dissolve in water – oil paint, for example, will not dissolve in water. The table lists some other solvents and the substances that are soluble in them.

→ Figure 10.5 Some substances that do not dissolve in water can be dissolved in these solvents.

Solvent	Soluble substances
acetone	nail polish, grease, glue
methylated spirits	grease, ballpoint ink
turpentine	oil paint, varnish
petrol	grease, oil
ethanol	flower fragrances (used to make perfume)

Experiment 10.2

Making a saturated solution

Aim

To find out how much sugar you can add to a beaker of water before it becomes saturated.

You will need:
- a beaker of water (at room temperature), a stirring rod, a teaspoon
- sugar

Method

Stir sugar into the beaker of water, one teaspoon at a time, until the solution is saturated.

Questions

1 How much sugar could your beaker of water hold before it became saturated?
2 Why did you stir the sugar-and-water mixture?
3 Is there anything you can do to make more sugar dissolve in this saturated solution? If you have ideas, think about how you could test them. Discuss this with your teacher.

↑ **Figure 10.6**
When no more solids will dissolve, a solution is saturated.

labels: stir, saturated solution, undissolved solid

Saturated solutions

What happens if you keep adding sugar to a solution of sugar and water? Can you just go on adding sugar to make the solution sweeter and sweeter?

The answer is that you cannot. There is a limit to how much sugar will dissolve in a certain amount of water. Eventually, a solution will get to a point at which no more solid will dissolve (Figure 10.6).

When no more solid will dissolve in a solution, the solution is **saturated**.

Factors that affect solubility

In Experiment 10.2, you stirred the sugar into the water. Stirring helps the solid and liquid mix faster and so the sugar dissolves faster.

If you crushed the sugar, it would also dissolve faster. Small pieces of a solid can dissolve faster than bigger pieces.

But the main way of speeding up dissolving is to heat up the liquid. Heat makes most solids dissolve faster.

Measuring solubility

The amount of a substance that dissolves in water is different at different temperatures. Scientists measure the **solubility** of a substance by working out how much of that substance will dissolve in 100 g of a solvent (usually water) before the solution becomes saturated.

Sugar has a solubility of 211 g in water at 25 °C. Salt has a solubility of 36 g in water at 25 °C.

The amount of a substance that dissolves in 100 g of water at different temperatures can be shown on a graph called a solubility curve. Figure 10.7 shows the solubility curves of salt (sodium chloride), potassium nitrate, lead nitrate and ammonia (NH_3).

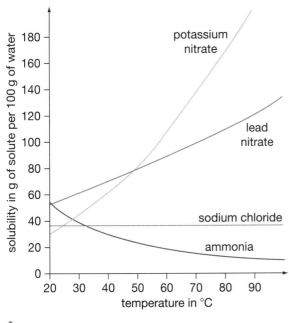

↑ **Figure 10.7** Solubility curves

If you study the graph in Figure 10.7, you should be able to see that:

- The solubility of salt, potassium nitrate and lead nitrate increases as the temperature increases. More of each substance will dissolve in warmer water. You can see this because the graph lines slope up to the right.
- The solubility of ammonia decreases as temperature increases. More ammonia will dissolve in cooler water. You can see this because the graph lines slopes down to the right.
- A steep curve, like that of potassium nitrate, shows a big increase in solubility at warmer temperatures. A fairly level curve, like that of salt (sodium chloride), shows hardly any increase in solubility with temperature.
- Where graphs cross each other, it means that the substances have the same solubility at that temperature.

 **Activity 10.2** **Solutions and solubility**

1 Match each word in column A to the correct definition in column B.

Column A	Column B
solution	the liquid in which things dissolve
soluble	the solid that dissolves
insoluble	unable to dissolve more solute
suspension	will dissolve
solvent	does not dissolve at all
solute	graph showing how much dissolves
saturated	mixture with large particles in it
solubility curve	mixture in which solid is dissolved

2 Explain the difference between a suspension and a solution. Give two examples of each.

3 How do you know when a solution is saturated?

4 What can you do to make substances dissolve faster?

5 Water is sometimes called the 'universal solvent'. Why do you think this is?

6 Name three other solvents. Name a substance that will dissolve in each.

7 What does it mean when we say that a substance has a solubility of 45 g at 25 °C?

Unit 3 Simple ways of separating mixtures

Many useful substances are found in nature as mixtures. For example, oxygen is found as a mixture with other gases in the Earth's atmosphere, salt is found in solution in sea water and gold is found mixed with rocks and water in rivers. Scientists have developed different methods of separating mixtures so that we can get the useful substances we need from them.

Decanting

When you are working with a mixture in suspension, you can separate it by **decanting**. To use this method you let the mixture settle, so the

solid falls to the bottom. Then you carefully pour the liquid off the top. This method is useful when you don't need an exact separation. For example, if you wanted to pour the water out of a pot of boiled vegetables, there would be some probably be some vegetable pieces in the water and some water would remain behind in the food.

↑ **Figure 10.8**
A colander is a type of sieve that allows you to separate solids and liquids in the kitchen.

Sieving and filtering

A **sieve** is a grid with holes in it which is used to separate mixtures by size. The colander in the photograph in Figure 10.9 has small holes in it. These holes are too small for the rice to fall through, but big enough for the water to run out.

Gardeners sometimes use a sieve made of mesh to sift stones from the soil. In large quarries, sieves are used to separate stones of different sizes.

Separating by size is often used in industry. In the food industry, eggs are sorted into different sizes, and small peas (and other vegetables) are separated from larger peas. In the mining industry, sieves are used to sort gravel containing diamonds, and in the fishing industry nets are made with small openings so that small fish can escape and only bigger fish are caught.

↑ **Figure 10.9** Filter coffee and tea are examples of filtration. The drink is the filtrate. The coffee grounds and tea leaves are the residue.

A **filter** is a very fine sieve. Filter paper is made from interlocking fibres. The filter paper allows liquid to pass through, but not larger solid particles. The substance (normally a liquid) that passes through the filter is called the **filtrate**. The substance that is left behind is called the **residue**.

Filter paper that you use at school is normally round and flat. You have to fold it to make a cone that can fit into a funnel when you use it. Figure 10.10 shows you how to do this.

Experiment 10.3

1

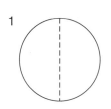

2

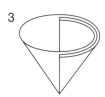

3

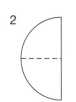

4

↑ **Figure 10.10** How to fold a filter paper

Decanting and filtering a suspension

Aim
To compare the results when you decant and when you filter a mixture of sand and water.

You will need:
- a funnel, filter paper and stirring rod ● a conical flask ● a beaker
- two test tubes half-filled with a mixture of water and sand (one teaspoon of sand per test tube)

Method
Let the mixture in one test tube settle. Then carefully decant (pour out) the water into the beaker.

Set up a funnel, filter paper and flask as shown in Figure 10.11. Filter the mixture from the second test tube.

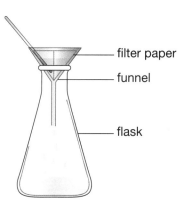

filter paper

funnel

flask

↑ **Figure 10.11** How to set up your filtering equipment

Questions
1 Is the liquid in the beaker after decanting clear or not?
2 Did decanting separate all the sand from the water?
3 Is the liquid in the flask (the filtrate) clear or not?
4 What is left on the filter paper after filtration?
5 Compare the two methods of separation.
   a) Which is better?
   b) Which is faster?
6 Do you think either of these methods will give you water that is clean and safe enough to drink? Explain why or why not.

Using magnets to separate mixtures

Substances that contain iron are attracted to magnets. This allows you to use magnets to separate mixtures of magnetic and non-magnetic substances – for example, a mixture of stones and nails, or a mixture of paper clips and buttons.

Magnetic separation is also used in industry. In the iron-mining industry, the rock that contains iron is broken up into small pieces. These pieces are placed on a large conveyor belt with a magnetic roller at one end. As the rock passes the roller, the pieces of rock that contain iron stay on the belt, and the other rocks fall off and are taken away. The magnetic pieces are collected and processed to get the iron from them. Magnetic separation is also used to purify kaolin clay, a material used to make white paper.

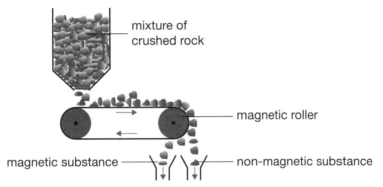

➡ Figure 10.12
A magnetic
separator

Evaporation

You know that water can change from a liquid to a gas (water vapour) at any temperature by evaporation. If there are solids dissolved in the water, they are left behind as the water evaporates. This makes evaporation a useful method of separating solids from solutions when you do not need to keep the liquid.

Case study: Using evaporation to get salt

In many places of the world, salt is made by evaporating sea water. Sea water is run into shallow pans in the sunshine. The heat from the Sun evaporates the water and the salt is left behind. This salt is then cleaned and purified so that people can safely eat it.

➡ Figure 10.13
Salt pans near the
coast in South
Africa

In some places, salt is mined from underground. The salt that is mined is mixed with sand and clay. The salt has to be separated from the sand and clay before we can use it. We can do this because salt is soluble but sand and clay are not. So, we can dissolve the salt and then filter the solution to remove the sand and clay. We can then evaporate the water from the salt solution to get the salt back.

Follow the steps in Figure 10.14 to see how this is done.

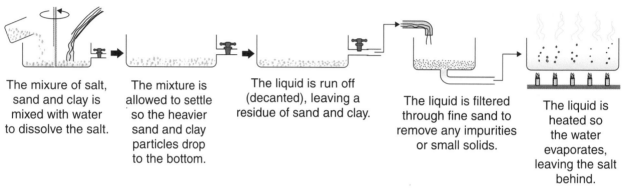

The mixure of salt, sand and clay is mixed with water to dissolve the salt.

The mixture is allowed to settle so the heavier sand and clay particles drop to the bottom.

The liquid is run off (decanted), leaving a residue of sand and clay.

The liquid is filtered through fine sand to remove any impurities or small solids.

The liquid is heated so the water evaporates, leaving the salt behind.

↑ **Figure 10.14** Using decanting, filtration and evaporation to get salt from a mixture.

Experiment 10.4

Separating salt from sand

Discuss how you could separate the salt from the sand in a dry mixture of the two solids.

Use the method you think is best to separate these two substances. You must have separate amounts of sand and salt at the end of your separation.

Activity 10.3 **Identifying methods of separation**

1 What methods of separation would you use to separate these mixtures?
a) scraps of iron from dirt in a workshop b) sugar from water
c) metal ball-bearings from a tub of oil   d) mud from water
e) small rocks from big rocks

2 Give an example of a mixture you could separate by:
a) magnetic separation b) decanting c) evaporation d) filtration
e) a combination of evaporation and filtration.

3 Why is filtration not useful for separating the solute and solvent in a solution?

4 People who live and work in large polluted cities sometimes wear masks over their noses and mouths. Why do you think they do this?

Unit 4   Other methods of separation

Distillation

Distillation is a method of separating that is used to get a pure solvent, such as water, from a solution. The process is similar to evaporation except that the liquid is not lost to the air.

In distillation, the liquid part of the mixture is evaporated. The gas (vapour) is collected and cooled in a different container to get the liquid back. The substance collected during distillation is called the **distillate**. The mixture that remains behind in the original container is called the **residue**.

Experiment
10.5

Distilling water

Aim
To separate and collect pure water from a salt solution.

You will need:
- a salt solution
- a beaker of cold water, flask, tube fitted with stoppers, test tube
- a Bunsen burner and tripod

Method
Pour the solution into the flask. Set up the equipment as shown in Figure 10.15.

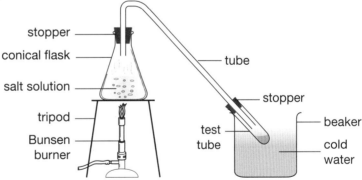

⬆ **Figure 10.15** How to set up your distillation equipment

Gently heat the mixture. Do not let the flask boil dry.

Questions
1 What collects in the test tube?
2 What remains in the flask?
3 How can you be sure that the substance in the test tube is water?

Chromatography

Different colours are mixed together to make inks and paints. These colours can be separated using a method called **chromatography**. 'Chromo' means colour.

Experiment 10.6

Splitting colours

Aim

To find out how many colours are mixed to make one colour in a felt-tip pen.

You will need:

- strips of thick paper
- coloured marker pens
- a large glass beaker
- water

Method

Choose four colours. Make a big dot of colour 2 cm from the bottom of each strip of paper. Write the name of each colour in pencil at the top of each strip.

Place 1 cm depth of water in the bottom of your glass beaker. Place the strips as shown in Figure 10.16. Leave the strips for 10 minutes.

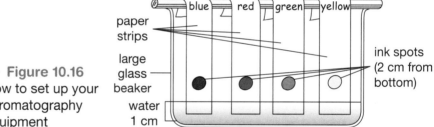

→ **Figure 10.16** How to set up your chromatography equipment

paper strips
large glass beaker
water 1 cm
ink spots (2 cm from bottom)

Questions

1 What did you observe?

2 Copy this table into your book and complete it to show your results.

Colour tested	Colours observed after it split

3 Discuss how the following factors might affect your results:
   a) the colour chosen
   b) the time that the strips are left in the water
   c) the type of marker used
   d) the distance of the dot from the water.

4 Choose one of the factors above and do a test to find out how it affects your results. Present your findings in diagrams and words.

Activity 10.4 **Applying your knowledge**

1 Imagine you were stranded on an island without fresh water. How could you use sea water to get fresh water?

2 How can you show that black ball-point ink is made from a number of different colours mixed together?

Chapter summary

✓ A pure substance is made up of only one type of substance.

✓ A mixture is made up of two or more substances.

✓ Solids, liquids and gases can combine in different ways to make different kinds of mixtures.

✓ Substances that dissolve in liquid are soluble. Substances that do not dissolve are insoluble.

✓ A solution is a mixture of liquid with another substance dissolved in it. The liquid is called the solvent, the dissolved substance is called the solute.

✓ A saturated solution cannot dissolve any more of a substance. If the conditions change, the solution may be able to dissolve more.

✓ Mixtures can be separated by decanting, sieving, filtering, magnetic separation and evaporation.

✓ Sometimes you need to use a combination of methods to separate a mixture.

✓ Distillation can recover pure liquids from a solution.

✓ Chromatography can be used to separate mixtures into colours.

Revision questions

1 Explain the difference between sieving and filtering.

2 What is decanting and when is it useful?

3 In a solution of coffee dissolved in hot water, what is the solute and what is the solvent?

4 In a mixture of chalk and water, what would you call the insoluble chalk that settles on the bottom of the container?

5 In a recipe, you are asked to dissolve three tablespoons of gravy powder in half a cup of water. But not all of the gravy powder dissolves. What could you do to make all the gravy powder dissolve?

6 Suggest two ways of separating each of the mixtures shown in Figure 10.17.

a) Oil and water   **b)** Salt and pepper

c) Staples and salt   **d)** Stainless steel and glass beads

↑ **Figure 10.17**

7 What is distilled water and how it is produced?

8 What is chromatography?

➜ **Figure 11.1**
What is making these girls bounce up and down?

The girls in the photograph are moving because of pushing and pulling forces. You can't see the forces, but you can see what they do. When the girls push down on the trampoline they stretch the springs that hold the mat and the mat sinks down a bit. The springs pull back and the mat tightens and pushes the girls upwards.

Forces act in many different ways. In this chapter, you are going to learn more about forces. You will see how forces can change the speed and direction of objects and how they can change the shape of objects. You will also measure forces using a special instrument called a newtonmeter.

As you work with forces and investigate their effects, you will:

- identify forces being used in different situations and classify them as pushing, pulling or twisting forces
- understand and give examples of the effects of forces
- read and interpret simple graphs of changes in speed
- measure the effects of a force using simple equipment.

Unit 1 Forces around us

Look at the pictures in Figure 11.2. Can you see that things are moving in each of these situations?

⬆ **Figure 11.2** What is causing the movement in each of these situations?

Each of these movements needed a push, a pull or a twist to get it started. In science, we use the word **force** to describe a pushing, pulling or twisting action. We use arrows to show the direction in which a force is working. You can see this in Figure 11.3.

⬆ **Figure 11.3** The arrows show the direction of each force.

Forces can be grouped into contact forces and non-contact forces. In a **contact force**, the object causing the force must touch the object or material that is feeling the force. Hitting a ball with a bat is a contact force. Pulling a rope is a contact force. In a **non-contact force**, the objects do not touch each other. A magnet can pull metal objects towards it without touching them. This is a good example of a non-contact force.

 **Activity 11.1** **Identifying and classifying forces**

1 Copy this table into your book. Look again at the pictures in Figure 11.2, and say which type of force is being used in each example – push, pull or twist. The first example has been filled in for you.

Action	Type of force
rolling dough	push
tightening a screw	
turning the pages of a book	
wringing out washing	
riding a bicycle	
pulling a horse-drawn cart	
opening a lock	
towing a car	

2 You use forces in your own life every day. Copy this table and fill in three examples of how you use each type of force.

Push	Pull	Twist

3 Draw simple diagrams with arrows to show the following forces:
   a) the push of a bat on a ball
   b) the pull of a hand on a dog leash
   c) the twist of a hand on a door handle.

Unit 2 The effects of a force

Finding out what forces do

You will need:
- a flat surface • a small ball

Method

Place the ball in the middle of the table so that it stays still in one place.

Try to make the ball move. Write down what you do to make it move.

How can you make the ball move in a different direction once it is moving? Write down what you do.

What can you do to make the ball stop once it is moving? Write down what you do.

Put the ball on the floor. Use a gentle force to make it move forward. Observe what happens to the ball as it moves.

Conclusion

Write a few sentences saying what this experiment has taught you about forces.

In science, when we use a force, we say that we are 'exerting' a force. When you exert a force, you can make objects move, change direction or stop moving. You cannot make things move, change their movement or stop them moving without exerting a force. The things that forces do are called the **effects** of the force.

- A force can make an object move.
- A force can make an object stop moving.
- A force can make things slow down or go faster – in other words, a force can change the speed of an object.
- A force can make a moving object change direction.
- A force can change the shape of some things.
- A pushing force can compress (squash) some things (such as bread dough or a spring).
- A pulling force can stretch some things (such as elastic bands or springs).

Some forces are much bigger than others. A big push will make the ball move further than a small push. A very hard pull will stretch an elastic band more than a small pull will.

Flexible and elastic objects

↑ **Figure 11.4**
Sandra is exerting a twisting force on this piece of rubber.

Flexible means able to bend easily. When you exert a force on a flexible object, such as a piece of rubber, you cause the object to bend and change its shape.

If Sandra stops exerting the twisting force on the piece of rubber in Figure 11.4, it will return to its original shape. We say that the rubber is **elastic** because it is able to return to its original shape. If you twist a thin piece of metal, it will bend and change shape because it is flexible. If you stop twisting, the metal will not go back to its original shape because it is not elastic.

Balanced and unbalanced forces

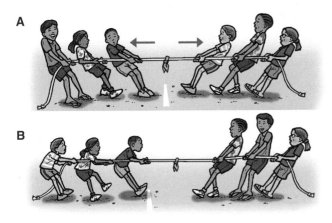

↑ Figure 11.5 Which forces are balanced?

In the example in Figure 11.5, both groups of children are pulling on the rope. This means that more than one force is acting on the rope at the same time. When the forces are equal (example A), they balance each other out and the forces have no effect on the object – in this case, the flag on the rope does not move. If one force is bigger than the other, then the force will act in the direction of the bigger force. In example B, the flag on the rope moves to the group on the right because they are exerting a bigger force than the group on the left. Unbalanced forces have an effect on the object.

Activity 11.2   Explaining the effects of forces

1 Read the seven effects of forces listed on page 130 again. For each effect, write down an example which shows this effect. For example, kicking a soccer ball makes it move.

2 Rubber is flexible. Give three more examples of flexible materials.

3 Very strong forces can make objects that are not flexible change shape. For example, when a car crashes into a wall, the car bonnet usually changes shape. Give two other examples to show how strong forces can make objects that are not flexible change shape.

↑ Figure 11.6

4 a) Look at Figure 11.6. Explain why the see-saw is not moving.
  b) What could these children do to make the see-saw move?

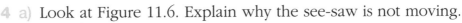

Unit 3 Changing speed and direction

Force can make movement faster or slower. Force can also change the direction of movement. In this unit, you are going to use graphs to understand how this happens.

Changing speed

→ **Figure 11.7**
Ling is pushing a trolley. His friend Anna joins him and helps him push.

The graph in Figure 11.8 is called a speed–time graph. Speed tells us how quickly or slowly an object is moving. This graph shows the speed of the trolley that Ling is pushing.

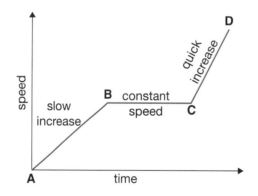

→ **Figure 11.8**
A speed–time graph for the trolley

- Before Ling pushed it, the trolley was stationary (not moving). It had no speed, so the line on the graph starts at zero.
- When Ling pushed it, the trolley began to move. It moved faster as he pushed. This is shown by line AB. Notice that the line goes upwards and to the right.
- After a short time, the trolley was going as fast as it could with Ling pushing. It went along at this constant speed for a few minutes. This is shown by line BC.
- Then Anna joined in. When she began to push, the force on the trolley increased and it began to move faster. This is shown by line CD. Notice that this line is steeper than line AB.

Changing direction

When a moving object, such as a ball, comes into contact with another object, such as a bat, it changes direction. First the ball is moving towards the bat, then it hits the bat and then it moves away from the bat. You can see this in action in games like tennis, cricket, badminton and squash. You can also see this in action if you throw a ball against a wall. Your throw pushes the ball towards the wall. When the ball hits the wall, the wall exerts a pushing force on the ball which causes it to bounce back and change direction.

Activity 11.3 **Reading graphs and diagrams**

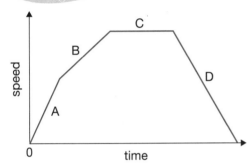

↑ Figure 11.9

1 Study the graph in Figure 11.9 carefully.

a) What type of graph is this?
b) Which lines on the graph show:
 • a quick increase in speed
 • a slow decrease in speed
 • a slow increase in speed
 • a constant speed?

2 Two girls are throwing a ball to each other. Which of the three graphs in Figure 11.10 best describes the speed of the ball as one girl throws it to the other?

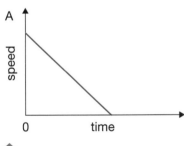
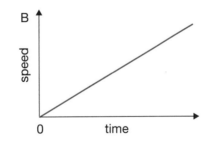
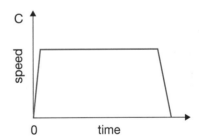

↑ Figure 11.10

3 Draw a diagram with force arrows to show what happens when a cricket ball hits a cricket bat.

Unit 4 Measuring force

Forces are measured in units called newtons (N). The newton is named after Sir Isaac Newton (1642–1727), an English scientist who developed scientific laws to explain how objects move. One newton is not a very big force. When you switch on a light you exert a force of about 5 newtons. When you open a can of cola you exert a force of about 20 newtons.

The instrument used to measure force is called a newtonmeter or force meter. Most newtonmeters use a spring to measure pull forces. Bathroom scales are a type of newtonmeter which is used to measure pushing forces. Figure 11.11 shows you three different examples of newtonmeters.

Experiment

Make your own newtonmeter

You will need:
- thick elastic bands ● large paper clips or a piece of wire
- an average sized apple ● a nail ● a strip of paper ● a ruler ● a pencil

Method

Set up your equipment as shown in Figure 11.12.

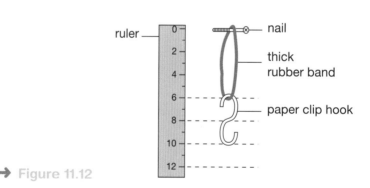

→ Figure 11.12

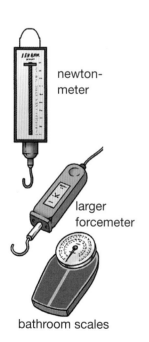

newton-meter

larger forcemeter

bathroom scales

Attach the apple to the hook of your newtonmeter.

The rubber band will stretch to 1 newton (the pulling force of an average apple). Use the ruler to mark off a scale from 1 to 12 newtons.

Collect different sized stones. Use your newtonmeter to find and record their weight in newtons.

↑ Figure 11.11 These devices all measure force in newtons. One kilogram on a set of bathroom scales is equivalent to 10 N.

Experiment

Who exerts the strongest pushing force?

Aim

To find out which pupil in your group can exert the strongest pushing force.

You will need:
- a set of bathroom scales

Method

Work in groups of five or six. Take turns to hold the bathroom scales in front of your chest, while standing against a wall.

Another pupil should then push as hard as possible against the scales, with both hands.

Repeat this three times. Record your results in a table like this one – remember to change kilograms to newtons. (1 kg = 10 N)

Pupil	Push 1	Push 2	Push 3	Average force of push (N)

Who exerted the strongest pushing force, on average?

Activity 11.4   **Measuring pulling forces**

1 Study Figure 11.13. The single elastic band in part A stretched 5 cm when a stone was attached to it.

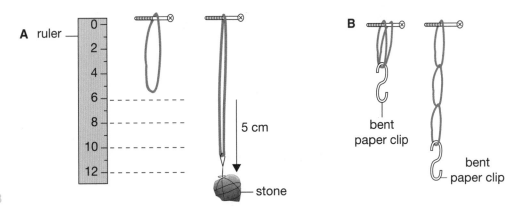

➡ Figure 11.13

a) Predict what will happen if you tie the same stone to each of the other sets of elastic bands (part B).

b) How could you test whether your predictions were accurate?

2 Write a set of instructions for using a newtonmeter to measure pulling forces.

Chapter summary

✓ A force is a push, pull or twist.

✓ Forces have an effect on objects. Forces can start movement, stop movement, change the speed and direction of movement, and change the shape of objects.

✓ Contact forces involve touch. Non-contact forces do not need touch.

✓ We can draw graphs to show how the speed of objects changes in response to different forces.

✓ Force can be measured in newtons (N) using a newtonmeter.

Revision questions

1 Heema is a taxi-driver. He says his life as a driver is easy because he doesn't need to use much force to drive. Study the photograph of the inside of Heema's taxi, in Figure 11.14.

a) Find out the names of the all the controls Heema uses as he drives. Write these as a list.

b) Next to each control on your list, write down whether Heema needs to use a push, pull or twist to operate the control.

c) Do you agree that Heema doesn't use force to drive? Give a reason for your answer.

↑ Figure 11.14

2 Explain what the following words mean:
  a) force b) compress c) flexible d) contact force e) newtons

3 You are trying to take a ball from a dog, who is holding it in his teeth.
  a) Draw a ball and add arrows to show the force exerted by the dog and the force exerted by yourself.
  b) Which of the forces is bigger?
  c) Are the forces balanced or not? Explain why.

4 Explain how you could measure the force needed to pull a brick up a ramp.

Chapter 12   Energy resources

⬆ Figure 12.1 **What do the battery and the snack bar have in common?**

The batteries and the snacks in the photograph can both supply energy. Batteries supply energy for various devices and snack bars supply energy for our bodies.

In this chapter you are going to learn more about energy and where it comes from. You will learn about sources of energy, including fuels, and about the different forms that energy can have. You will also learn how energy can change from one form to another. You will investigate heat energy to see how it can be transferred from one object to another and you will see how energy is wasted in the process.

As you work with energy and learn more about it, you will:

- understand what the word 'energy' means in science
- give examples of energy sources and classify them as renewable or non-renewable
- give examples of fuels and their uses
- list different forms of energy and give examples of each form
- draw diagrams to show how energy can change from one form to another
- investigate heat transfer by conduction, convection and radiation
- distinguish between good and bad conductors of heat
- calculate the percentage of heat lost in different energy transfers.

Unit 1 What is energy and where does it come from?

Everything we do requires energy – when we run, eat, talk, think, sleep or study, we are using energy. All things that work use energy – cars, trains, bicycles, ploughs and so on. In science, **work** is done when a force causes something to move. When work is done, energy is changed from one type to another. This is really quite simple. Think about a car. The engine of the car burns petrol, which is a form of stored energy. The burning petrol makes the pistons in the engine move and these make the car move forward.

Without energy there could be no movement. Without movement there could be no life on Earth. Where does all this energy come from?

All energy comes from a source. We get energy from two types of sources: renewable sources and non-renewable or finite sources.

A **renewable** source of energy is one that can be replaced quickly and easily. Humans and animals get energy from food. Food is a renewable source of energy because we can grow or buy more food when our supply is used up. Wood is also an example of a renewable source of energy because trees that have been felled can be replaced by replanting.

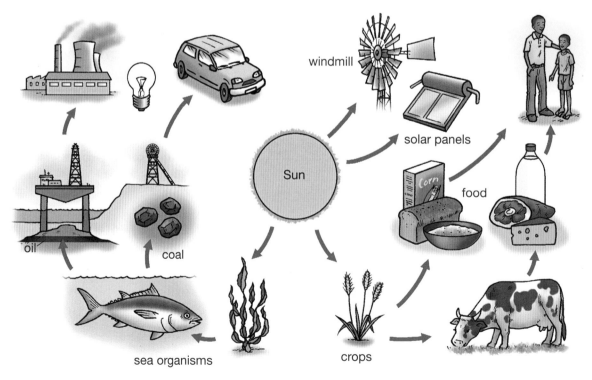

windmill

solar panels

food

oil

coal

sea organisms

crops

Sun

↑ Figure 12.2 There could be no life on Earth without energy from the Sun.

A **non-renewable** or finite source of energy cannot be replaced once it has been used up. Coal and oil are examples of non-renewable energy sources. If we use up all the coal in the world, we will not be able to make or grow more coal. It takes millions of years for coal to form naturally so we say it is non-renewable.

If you study Figure 12.2 you will the see that the Sun is the original source of most energy on Earth. All the other sources of energy shown in the diagram can be traced back to the Sun. The amount of energy that Earth receives from the Sun is equivalent to the energy we would get if we burned 6 million tonnes of coal every second!

Fuels

In science, we use the word **fuel** to describe a substance that can be burned to release energy. Wood, coal, petrol, oil and paraffin are all fuels. When we burn these fuels they give us energy in the form of heat and light.

Fuels like coal and oil were formed millions of years ago from the bodies of dead plants and animals. The hard remains of a plant or animal are called a fossil, so we say that coal and oil are fossil fuels.

Identifying sources of energy

1 Look at the pictures in Figure 12.3 carefully.

A  B  C  D  E

↑ Figure 12.3

a) What is energy being used for in each picture?

b) Name the sources of energy being used in each picture.

2 Make a list of all the fuels you use in your daily life. Does each of them come from a renewable or a non-renewable source?

Unit 2 Different forms of energy

Figure 12.4 shows some of the different forms of energy that we use on a daily basis. There are many different forms of energy. Remember, as you read about each form, that energy can change from one form to another.

→ Figure 12.4
Different forms of energy

Electrical energy

Electric currents in wires carry electrical energy from power stations to our homes. Electric motors also use electrical energy to do work. The energy carried to a motor by an electric current may come from the stored chemical energy in a battery.

Chemical energy

The energy stored in food is called chemical energy. This is because food contains many different chemicals. When we eat and digest food, our bodies break it down into its chemical components. It is these chemicals that get into our cells and provide us with energy.

Fuels like petrol and oil also contain chemical energy. When these fuels burn in oxygen, their chemical energy is changed into heat energy.

Light and sound energy

Light energy and sound energy are forms of energy that travel as moving vibrations called waves. Light waves, radio waves, microwaves and X-rays are all types of **electromagnetic waves** which carry different amounts of energy.

Heat (thermal) energy

Heat is a form of energy because it can be used to do work. For example, the heat in the picture in Figure 12.4 is being used to cook food. The heat causes the particles in the food to vibrate (move) quickly.

Nuclear energy

Atoms are tiny pieces of matter that store large amounts of energy. Nuclear energy is released when atoms are split apart (fission) or combined (fusion). Nuclear power stations use energy from atoms to produce electricity.

Stored energy

When energy is contained in an object such as a battery, or in a substance such as oil, we call it stored energy. There are three forms of stored energy: chemical energy, nuclear energy and potential energy.

Objects which have been pushed or pulled into a position where they can do work have potential energy. When a spring is pushed down it has potential energy. When an elastic band is stretched it has potential energy. When a stone, or water, is kept above the ground they have potential energy. The potential energy is released when the object is released and allowed to go back to its original shape or when it is allowed to fall back to Earth (gravitational potential energy).

Kinetic energy

Kinetic energy means moving energy. Things that are moving have kinetic energy. Figure 12.6 shows you some examples of kinetic energy.

dropped stone

stretched band

squashed spring

↑ Figure 12.5
These objects all have potential energy.

→ Figure 12.6
These things all have kinetic energy.

  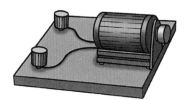

windmill   waterfall   dynamo

Activity 12.2   Describing forms of energy

1 a) Write down five forms of energy.
   b) Give three examples of things which have each form of energy.

2 Electricity is an important source of energy in the modern world.
   a) List ten ways in which you make use of electrical energy in your daily life.
   b) Describe how your life would change if there was no electricity.

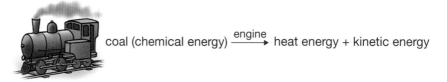

Unit 3 Energy changes

Scientists have discovered that energy can never be made or destroyed. This is called the Law of Conservation of Energy. This law means that energy can be changed from one form to another, but it is never used up, and the total amount of energy stays the same. Each time the energy is changed from one form to another, some energy ends up as heat energy which spreads out around the Earth.

Energy can only be used to do work when it changes from one form to another. For example, a gas stove uses gas. The gas has stored chemical energy. Burning the gas makes heat. The stored energy in the gas is changed into heat energy (the hot flame) and some is changed into light energy (the light from the flame). We use simple word equations to show these energy changes:

$$\text{gas (chemical energy)} \xrightarrow{\text{gas stove}} \text{heat energy} + \text{light energy}$$

Figure 12.7 shows you some energy changes.

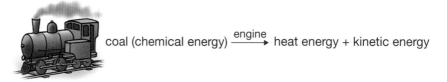
$$\text{coal (chemical energy)} \xrightarrow{\text{engine}} \text{heat energy} + \text{kinetic energy}$$

$$\text{batteries (chemical energy)} \xrightarrow{\text{torch}} \text{light} + \text{heat}$$

$$\text{batteries (chemical energy)} \xrightarrow{\text{radio}} \text{sound}$$

$$\text{potential energy} \xrightarrow{\text{yoyo}} \text{kinetic energy (movement)}$$

$$\text{potential energy} \xrightarrow{\text{toy}} \text{kinetic energy (movement)} + \text{sound}$$

➜ Figure 12.7
Energy changes

Sometimes it is difficult to tell what energy changes have taken place (see Figure 12.8). This may be because there is more than one source of energy to start with or because you cannot tell what the source of energy is. Sometimes, the energy is changed to one form and then to another, or many different changes take place at once.

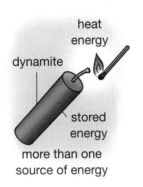

heat
energy

dynamite

stored
energy

more than one
source of energy

radiant
energy

stored chemical
energy

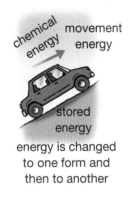

chemical
energy

movement
energy

stored
energy

energy is changed
to one form and
then to another

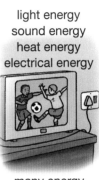

light energy
sound energy
heat energy
electrical energy

many energy
changes happen
at once

➡ **Figure 12.8**
Sometimes energy
changes are difficult
to work out.

Experiment

Changing energy

Work in groups. Make a water wheel like the one in Figure 12.9.

Pour water onto your water wheel to use it. Observe the energy changes that take place.

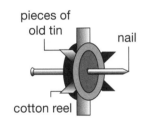

pieces of
old tin

nail

cotton reel

⬆ Figure 12.9

Questions
1 What energy changes take place when you use your water wheel?
2 How are machines like this used in real life?

Activity 12.3 · Giving examples of energy changes

1 Mr Jones is showing his class how energy changes. He uses tongs to hold a piece of steel wool in a flame. After a few minutes the steel wool glows red, and then sparks start to fly off it. What energy changes take place in this situation?

2 Think about a famous sportsperson.
   a) How does a sportsperson get energy to compete in races or events?
   b) What energy changes take place when a sportsperson moves?

3 Give an example of each of these energy changes:
   a) kinetic → electrical b) kinetic → heat c) potential → kinetic
   d) chemical → sound e) chemical → movement and sound

Unit 4 Transferring heat energy

When heat energy is moved from one thing to another – for example, from a hot cup to your hands – we say the energy has been **transferred**.

Heat energy can be transferred in three ways – by conduction, convection and radiation.

- **Conduction** is the transfer of heat energy through a solid material.
- **Convection** is the transfer of heat energy in moving liquid or air (gas).
- **Radiation** is the transfer of heat waves through space. The Sun's energy reaches Earth by radiation.

Heat energy always passes from a hotter object to a cooler object.

Conduction

Experiment

Predicting results

You will need:
- a beaker half full of just-boiled water
- a long piece of wire
- a long thin twig

Method

Put the wire and the twig in the hot water.

Hold the end of the wire with one hand and the end of the twig with the other.

Predict which will get hot first.

Keep holding the wire and the twig to test your prediction.

Safety note

Work carefully so that you do not burn yourself or anyone else.

Heat spreads through solids, like the wire, by conduction. Solids which can transfer heat easily are called **conductors**. Metal is a good conductor of heat. Wood is a poor conductor of heat. Poor conductors are also called **insulators**. We use insulators to stop heat being transferred from one object to another. For example, metal cooking pots often have handles made of wood or plastic to stop the heat being transferred from the pot to your hands.

Comparing insulating materials

Work with a partner.

Collect samples of paper cups, plastic cups and polystyrene cups used at fast food outlets.

Predict which cup will keep a hot drink warmest for the longest time.

Design and carry out a test to see how accurate your prediction was.

Convection

Making currents

You will need:
● a beaker ● water ● sand ● a burner ● matches

Method
Put some water into the beaker. Drop some sand into the water.
Heat the beaker over a low flame.

Questions
1 What happens to the sand?
2 Try to explain this.

When heat energy spreads through a liquid or gas, we say the heat has been transferred by convection.

When you heat water, the warm water becomes less dense and it rises to the surface. The colder water then sinks to fill the space left by the warm water. This causes movements called convection currents. If you heat a liquid like you did in Experiment 12.4 you can observe convection currents in action.

 **Investigating conduction and convection**

1 List five examples of insulators in your own home. Next to each one, say what it is used for.

2 Explain why the end of a teaspoon gets hot if you leave it in a cup of hot tea.

Radiation

Heat and light energy are transferred by radiation through space. This is how heat and light from the Sun reach the Earth.

Rays from the Sun pass through space and they give off some of their heat when they hit an obstacle. For example, when the Sun's rays hit the ground, they give off heat.

Some objects absorb radiant heat and become hotter themselves. Other objects reflect radiant heat so that it bounces off them back into space. Light and heat energy are easily reflected by light-coloured, shiny surfaces but not so well reflected by darker, dull or rough surfaces. This is why mirrors are smooth and light coloured. It is also why white or silver cars are cooler in summer than darker-coloured cars – sunlight is reflected off the surface of white and silver cars, but it is absorbed by the darker colours.

Experiment

Absorbing radiant heat

Aim

To test materials of different colours to see which ones absorb most heat.

You will need:
- three clear plastic bottles ● a black plastic bag
- a white plastic bag ● tin foil ● a thermometer

Method

Fill the three bottles with water that is the same temperature.

Cover one bottle with black plastic, one with white plastic and one with foil.

Put the three bottles in the same place in bright sunlight for half an hour.

Predict which bottle will have the warmest water after half an hour. Give a reason for your prediction.

Measure the temperature of the water in each bottle after half an hour. Record your results.

Questions

1 Which bottle had the warmest water?
2 Which bottle had the coolest water?
3 What can you conclude from this experiment?

Heat transfer wastes energy

When energy is transferred, some energy is wasted (lost). Energy which is wasted does not do any work. For example, in a normal incandescent light bulb, only around 10% of the electrical energy transferred to the bulb is changed into light energy. The other 90% is wasted because it is changed into heat energy which is transferred to the air around the light bulb. This heat energy does not do any work, so it is called wasted energy.

No machine can change 100% of the energy it receives to useful energy. A car, for example, loses heat and sound energy as it moves.

 **Activity 12.5** **Applying your knowledge**

1 A dark green car, a silver car and a white car are parked outside in the sunshine with all their doors and windows closed.
   a) Which car will be hottest inside after an hour? Why?
   b) Which car will be coolest inside after an hour? Why?
   c) How would these results be different if the dark green car's windows were all open? Why?
   d) Why is it dangerous to leave children or pets locked in a car?

2 The table shows you how much energy is changed to useful energy by different devices. Copy the table and fill in the missing information.

Device	Source of energy	Energy changed into useful energy (%)	Energy wasted (%)	Energy changes that take place
gas stove	natural gas	80		chemical → heat chemical → light
paraffin stove	paraffin	60		
wood stove		60		
fireplace		14		
diesel engine		38		
petrol engine		25		
nuclear power station		30		
solar panel		10		
battery		90		

Chapter summary

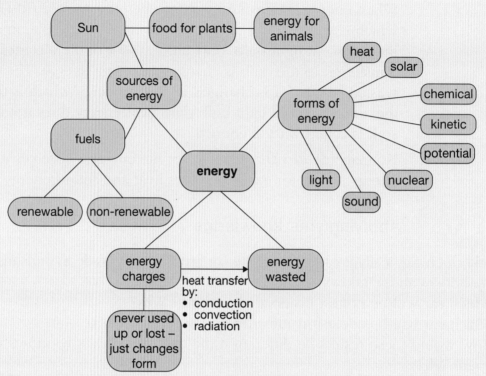

⬆ Figure 12.10 This is a summary of what you have learned about energy.

Revision questions

1 Name four renewable sources of energy.

2 Explain why coal and oil are not renewable sources of energy.

3 Make a table to show at least five forms of energy and give two examples of each form.

4 Why does a car stop when it runs out of fuel?

5 List five good conductors of heat and five good insulators.

6 How does heat travel from a stove plate to food in a pot?

7 How can you cause a convection current in a basin of cold water?

8 Why do you get sunburned when you stay out in the sunshine?

9 Why does chocolate melt if you leave it out in the sunshine?

10 Which material is better for building a shelter: wood or corrugated iron? Give a reason for your answer.

Chapter 13 | Electrical circuits

→ **Figure 13.1**
The lights in this city can only come on if they are connected to an electricity supply and switched on.

How does electricity work? What happens when you switch on an appliance? In this chapter, you are going to investigate how electricity works using simple circuits. You will build circuits and use symbols to draw circuits to find out more about switches, components, conductors and insulators.

As you complete the activities on circuits you will:

- describe a complete circuit and explain why it must be complete for energy to flow
- recognise and draw symbols to show components of a simple circuit
- test materials to find out which are conductors and which are insulators
- investigate the effects of adding cells, lamps, buzzers and switches to a circuit.

Unit 1 Electrical energy and circuits

In cities today, electricity is available when you turn on a switch. But moving electricity along wires to provide power for lights and machinery has only been possible in the last 150 years (since the late 1800s). Today, electricity is one of the most useful sources of energy for lights, to provide power for industry and even in motor cars.

Electric current

↑ **Figure 13.2** Modern batteries have a positive and a negative terminal. If you do not connect them to your appliances in the correct way, no current will flow.

When you switch on lights or an electrical appliance, electricity flows along the pathway made by the wires. This allows us to send electricity from its source (at the power station) to places where it can be used. The electricity that flows along the wires is called an **electric current**.

You will learn more about electric currents once you have learned about atoms and electrons.

When scientists first worked with electric current, they believed that it flowed from the positive end of a circuit to the negative end of the circuit. If you look at modern batteries like the ones in Figure 13.2 you will see that they have a positive (+) and a negative (−) end or terminal.

Today scientists know that current actually flows from the negative terminal to the positive terminal. But, when you work with circuits, you still show a flow from positive to negative. This is called the conventional direction of current.

Note about cells and batteries

In science, the correct name for a single source of electrical energy is 'cell', but we normally call this a battery. The batteries that you buy in the shops are mostly single cells. In science, a battery is made of more than one cell. A large car battery *is* a battery.

Electrical circuits

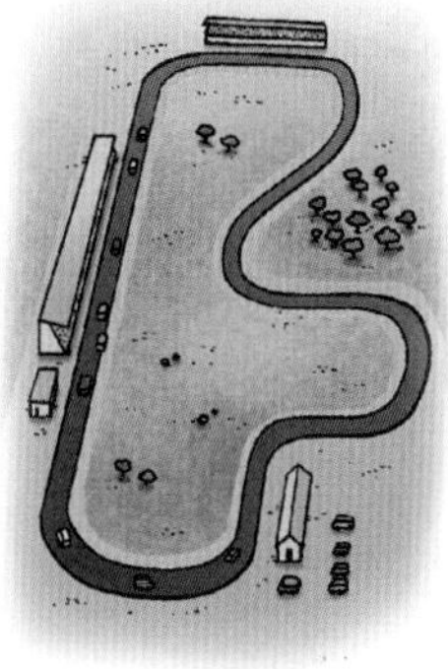

↑ **Figure 13.3** Cars race round this circuit. They start and end at the same place.

Formula One racing cars race round a track called a circuit. They start and finish at the same place on the track. Each time they go round the track we say they have made a circuit. A circuit is a pathway or track that goes back to its starting point.

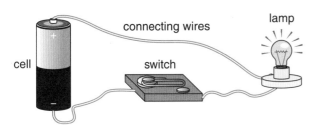

↑ Figure 13.4
A simple electric circuit

Figure 13.4 shows you an **electrical circuit**. When the circuit is connected and switched on, the current flows from its source at the cell around the circuit. The wires, the cell, the lamp and the switch are called the **components** (parts) of the circuit.

Making a circuit

All electrical circuits need three parts so that they can work:

- A source of energy – such as a battery or power supply. At school, you will use a cell or a battery as a source of energy for your circuit. At home, the source of energy is usually the mains supply which is connected to a power station. Circuits can also use solar cells or generators.
- Connections – the connections allow the current to flow from the source and around the circuit. The connecting material must be a conductor – it must allow current to flow. Most conductors in a circuit are made from metals. You will probably use copper wire as a conductor in your circuits at school but other metals can also be used.
- A resistor or output device that changes electrical energy into another form of energy. A lamp is an example of a resistor that you will use at school. At home, there are many different output devices attached to the electrical circuit: lights, refrigerators, computers, air-conditioners and televisions are all examples of output devices or resistors.

Switches and controls

Most circuits also have a switch or other device to turn the current on and off. The switch closes the circuit when you switch it on and this allows the current to flow. When you switch it off, the circuit is broken and the current cannot flow around it.

Safety note

Never do circuit experiments with electricity from the mains socket. A mains socket is the source of a large amount of electricity which will give you a serious electrical shock and which can kill you.

 **Activity 13.1** | **Understanding circuits**

1 What is an electrical circuit?

2 List three examples of electrical circuits in your daily life.

3 Where does the energy come from in a circuit?

4 What is a conductor?

5 Why do you need conductors in a circuit?

6 What are switches used for in a circuit?

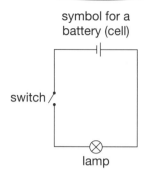

symbol for a
battery (cell)

switch

lamp

↑ **Figure 13.5**
A circuit diagram.
Can you recognise
the different
components?

Unit 2 Circuit components and symbols

Scientists use symbols to show the components and structure of a circuit. The circuit on page 151 looks like Figure 13.5 when it is drawn as a circuit diagram using symbols.

Before you can draw circuits you need to know what symbols to use (Figure 13.6). The symbols for parts of a circuit are standard symbols, used by scientists all over the world.

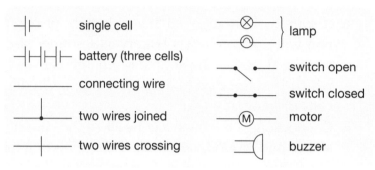

⊣⊢ single cell	⊗ ⊘ } lamp
⊣⊢⊣⊢⊣⊢ battery (three cells)	— switch open
—— connecting wire	— switch closed
⊥ two wires joined	Ⓜ motor
+ two wires crossing	◁ buzzer

↑ **Figure 13.6** Symbols used to show components of a circuit

Connecting lamps to a circuit

Most lamps have two parts: a small bulb, and a holder into which the bulb is screwed. The holder normally has two terminals which can be connected to the wires in a circuit. Be careful when you connect lamps to a circuit, and always screw the bulb into the holder before connecting it to the circuit.

The wire (filament) inside a lamp will burn if the electric current is too strong. When that happens the bulb has 'blown' and it cannot be used any more.

Switches

↑ **Figure 13.7**
These are all types
of rocker switches.

There are many different kinds of switches. When you work with simple circuits you will use simple on–off or rocker switches. When the switch is closed, the metal parts make contact, the circuit is completed and the current can flow through it. When the switch is open, the circuit is broken and no current can flow. Figure 13.7 shows you some of the rocker switches you might use at school.

Experiment 13.1

Making a switch

You will need:
- a cell, lamp and conducting wire
- paper clips and drawing pins

Method

Build a circuit like the one in Figure 13.8, to test the switches you make.

Design and make a switch using drawing pins and a plain metal paper clip. Make another switch using a plastic-coated paper clip.

Explain any differences that you observe.

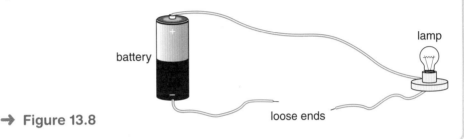

→ **Figure 13.8**

Activity 13.2 Drawing circuits

1 Use the symbols that you have learned in this unit to draw circuit diagrams of the circuits in Figure 13.9.

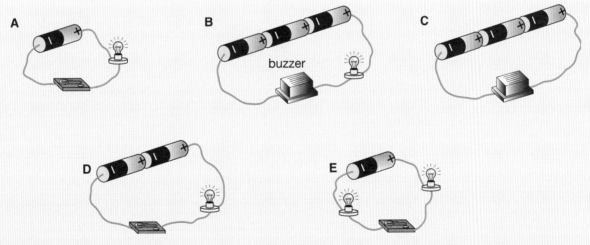

↑ **Figure 13.9**

2 How does a buzzer change electrical energy to another form of energy?

Unit 3 Conductors and insulators

You have learned that electric current flows through solid materials called conductors. All materials that conduct electricity are called conductors.

Some materials do not allow electricity to flow through them easily. These materials are called insulators.

We say that insulators are resistant to the flow of current. Conductors have very little **resistance**, so current flows through them easily.

You are going to carry out an experiment to find out which substances conduct electricity and which substances do not conduct electricity.

Experiment 13.2

Investigating conductors and insulators

Aim

To test different materials to find out if they conduct electricity.

You will need:
- a test circuit set up as shown in Figure 13.10
- a range of materials for testing, such as metals, plastics, wood, rubber, glass, cardboard, cloth, pencil lead, plant materials and any other substances you can find

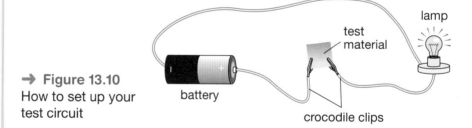

→ **Figure 13.10**
How to set up your test circuit

Method

1 Touch the crocodile clips together to check that the circuit is working and that the lamp lights up. If it doesn't, check that the lamp is not blown and that all the connections are good.
2 Test each material by attaching it to the circuit with the crocodile clips. If the lamp lights up, the material is a conductor.
3 Draw a table like this one in your book. Record your results by ticking the correct column for each material.

Material	Conductor	Insulator

Birds and electricity cables

The thick wires used to carry electricity from power stations to cities and homes carry enough electricity to kill a person. But birds can sit on these wires without getting a shock.

Birds do not get a shock when they sit on electrical cables as long as they only touch one wire. If they only touch one wire they are not part of the circuit so electricity cannot flow through them.

If the bird touches another wire, the electricity pylon or the ground at the same time, it will complete the circuit. The current will flow through the bird's body and the electric shock will kill it.

Large birds like storks and vultures are often killed by electric cables. This is because their wingspan is so big that they sometimes touch two cables as they try to land on one of them.

↑ **Figure 13.11**

Activity 13.3 Sorting conductors and insulators

1 Sort the materials in the box into conductors and insulators. (If you are not sure, you could test the material using your test circuit.)

> copper wire   eraser   plastic   elastic band   stick   coin   paper   wooden ruler
> wool   metal scissors   glass   paper clip   string   shoe laces

2 Copy and complete these sentences:
   a) Electrical wires are made from _____ because it has low resistance to current.
   b) Electrical wires are often covered with plastic because it is a good _____.
   c) Electrical wires must be insulated to reduce the risk of electric _____.

3 Explain why electricians working on electric wires wear thick rubber gloves and shoes with thick rubber soles.

Chapter summary

✓ An electric circuit is a pathway for electric current.

✓ All circuits must have an energy source, conductors and resistors (output devices). Most also have some sort of switch.

✓ When we draw electric circuits we use internationally accepted symbols to show the different components.

✓ Circuit diagrams are always drawn with straight lines and right angles, which makes them look like boxes.

✓ Materials that allow electricity to flow through them are called conductors.

✓ Materials that resist the flow of electricity are called insulators.

✓ Most electrical wires are insulated to prevent electric shocks.

Revision questions

1 Write down the names of the components labelled A to D in the circuit in Figure 13.12.

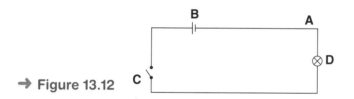

➡ Figure 13.12

2 Look at the three circuits in Figure 13.13.

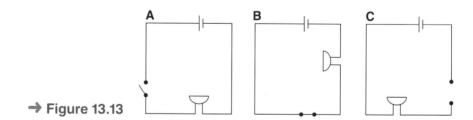

➡ Figure 13.13

a) The buzzer will only make a noise in one of these circuits. Which one?

b) Why would the buzzer not make a noise in the other two circuits?

3 Does each of these materials conduct electricity?
a) wood   b) steel   c) silver   d) rubber   e) aluminium foil

4 Name four materials that can act as insulators.

 The Earth in space

↑ **Figure 14.1**
This girl is
studying space.

Many of the things you know about the Earth are things you can see
and make sense of in your daily life. It is quite easy to observe things
that happen around us.

In this chapter, you are going to learn about things that you cannot
easily see or observe. You are going to learn about the Earth as a
moving part of a larger system in space. You are also going to learn
what scientists do to find out about the Earth in space. The study of
space is also called **astronomy**.

To understand about the Earth and its position in space, you need to
be able to:

- make sense of diagrams and read information to learn about
  different theories in astronomy at different times in history
- carry out observations to record the phases of the Moon
- complete diagrams to show how a solar eclipse and a lunar eclipse
  occur
- answer questions about the Earth's atmosphere
- read tables, graphs and maps to prepare a presentation about the
  ways in which humans can affect the atmosphere.

Unit 1 Finding out about space

People have known for thousands of years that there is some movement in space. People saw the Sun rising in the morning and setting at night and they knew that stars appeared at night and disappeared in the morning. Different cultures and peoples explained the things they observed in ways that made sense to them. Many of these explanations were closely linked to religious beliefs.

What did people believe long ago?

↑ **Figure 14.2** An African story explains that the stars are actually sunlight shining through holes in a blanket. They believed the gods threw the blanket over the Sun at night to let the people and animals sleep.

↑ **Figure 14.3** The Greeks believed the stars and planets were attached to big crystal spheres that were made to turn by machines. They thought if they could break the spheres they'd be able to see the machines.

↑ **Figure 14.4** By the second century, many people believed the Earth was the centre of the Universe. They believed the Sun, Moon, stars and planets all moved around the Earth. The Christian church supported this idea.

↑ **Figure 14.5** The invention of the telescope allowed scientists to prove that the Earth, Moon, stars and planets all move around the Sun.

Improved equipment, advancements in mathematics and the development of powerful computers have all allowed scientists to find out more about space. Today we can even send robots to explore the surfaces of other planets. These robots send pictures and information back to computers on Earth.

Case study: A modern astronomer

As a young boy, Thebe Medupe lived in a village in southern Africa. His family was poor and the village did not have electricity. Thebe remembers that he used to sit outside round the fire and listen to his elders tell traditional stories. It was there that his love of the sky began.

At high school, Thebe loved science and mathematics. When he was 13, he learned about Halley's Comet and decided to build his own telescope. He used a cardboard tube and lens which he was given were donated by the school laboratory technician. Thebe took his telescope outside at night and discovered that it worked! He spent hours looking at mountains, plains and craters on the Moon. You can see Thebe with his telescope in Figure 14.6.

→ **Figure 14.6**
Thebe Medupe

Thebe won a science award at school and then he decided to study the stars at university. Today he has a doctorate in astrophysics and is a lecturer and researcher. He has also become famous as the star of a film called *Cosmic Africa*.

Thebe explains that he found it very disturbing that his friends and other pupils seemed to think that the African way of life was not important enough to learn about. He disagreed and has travelled around Africa to learn from local people and to share his ideas. He discovered that indigenous people in Africa have very deep connections with space. It is this journey and the rich stories he gathered that form the backbone of the international film in which he starred.

Activity 14.1 **Understanding how ideas can change**

1 Read the information on pages 158 and 159, including the case study.

   a) The diary entry below shows what an astronomer living in the 1500s saw when he was studying the sky one night.

> *12 November*
>
> *It is bitterly cold outside. I climbed onto the roof platform to see whether I could get a good view of the planet Mars. The Full Moon made the sky very light, so it was difficult to see much detail. There was also low cloud that made it difficult to see.*
> *I wish my eyesight was better, and then I'd be able to see more. Once I'd finished my observations, I made notes and sketches in my journal.*

     Imagine you are a modern astronomer. Write a diary entry explaining how you study the night sky.

   b) Write two sentences explaining why ancient and modern astronomers had such different ideas about the Universe.

2 Imagine that Dr Thebe Medupe has been invited to talk to your class. Work in groups to make up a list of five questions that you would like to ask him.

3 Is it easy to build a telescope? Find out how a telescope works and write instructions for making a simple one.

Unit 2 The Solar System

The word 'solar' means 'to do with the Sun'. It comes from the Latin word 'sol', which means sun.

The Solar System is the name given to our Sun and all the things that move around it. These include:

- nine planets and planetoids, including Earth, and their moons
- asteroids
- comets
- dust and gas.

Note about Pluto
Astronomers have decided that the planet Pluto is not a real planet. It was reclassified in 2006 and it is now called a planetoid.

The Sun is the most important object in our Solar System but it is really just a star. Stars are very large, spinning balls of hot gas. The Sun looks really big compared to other stars, but that is because it is much closer to Earth than they are.

Planets are different from stars. Stars give out light, but planets do not give out light of their own. We can see planets in the sky because they reflect light from the Sun. All the planets in the Solar System move, or revolve, around the Sun in a regular pattern, called an **orbit**. Each planet takes a different amount of time to complete an orbit.

Activity 14.2 Finding out about the planets

1 Study Figure 14.7 and read the information about the planets on pages 163 and 164.
   a) List the planets in order from the one closest to the Sun to the one furthest away.
   b) List the planets in order of size from biggest to smallest.
   c) Which planets have rings around them?
   d) Which planets do not have any moons?
   e) Between which planets do you find the asteroid belt?

2 The diameter of the Sun is ten times bigger than the diameter of the largest planet. Use a pair of compasses to draw a scale diagram which shows this difference in size.

⬆ **Figure 14.7** The Solar System

Remember that the Solar System is only a small part of space. The Solar System is like a little suburb in a very big, ever-growing city called the Universe. The Universe is the name given to everything that exists – from our Earth to the furthest stars and all the space in between.

Planet facts

Planet and appearance	Average distance from Sun in millions of km	Diameter in km (approx)	Length of revolution (one year) in Earth days (approx)	Surface conditions	Number of moons known in 2007 (changes all the time)
Mercury looks orange	58	4 880	88	350 °C day, 180 °C night, no atmosphere, no water	0
Venus appears bright yellow	108	12 100	225	500 °C, atmosphere is 98% carbon dioxide, thick clouds	0
Earth looks blue from space	150	12 682	365	22 °C, oxygen rich atmosphere, water present	1
Mars seen as a red dot	228	6 720	687	−30 °C, thin atmosphere of carbon dioxide, ice at poles, serious dust storms	2
Jupiter has a red dot in a yellow background	778	141 920	4 344	−150 °C, thick almost liquid hydrogen atmosphere, strong magnetic field	60
Saturn looks orange	1 427	120 000	10 755	−180 °C, gassy surface with ammonia clouds, very strong winds, thousands of rings	31

Planet and appearance	Average distance from Sun in millions of km	Diameter in km (approx)	Length of revolution (one year) in Earth days (approx)	Surface conditions	Number of moons known in 2007 (changes all the time)
Uranus appears blue	2 871	51 160	30 664	−210 °C, strong winds on surface, 11 almost invisible rings, poles are the warmest places on the planet	27
Neptune can't be seen clearly	4 497	49 860	60 148	−220 °C, atmosphere of hydrogen and methane, very strong winds and hydrocarbon snow	13
Pluto can't be seen clearly	5 914	2 300	90 717	−230 °C (estimate), thick atmosphere of nitrogen and methane, not much known about surface	1

Remember that information about planets and space changes all the time. When scientists discover new things, their results are published and they have to be checked and approved by other scientists before they are accepted.

Activity 14.3   Comparing different planets

1 What three conditions do you think humans and other animals need so that they can survive on a planet?

2 Write down three reasons why Earth is the only planet suitable for human life.

3 Choose one other planet. Explain why it would be impossible for humans to live naturally on the planet you have chosen.

4 If there was human life on one of the other planets, which planet would be most likely? Why?

Activity 14.4 Making a model of the Solar System

You will need:

- cardboard
- compasses
- scissors
- glue
- ruler
- crayons or paints
- a wall big enough to display your work

You are going to work in groups to make a scale model of the Solar System. The sizes have been worked out using two different scales:

- Pluto (the furthest planet) is 10 metres away from the Sun.
- Jupiter (the largest planet) is 30 centimetres in diameter.

	Mercury	Venus	Earth	Mars	Jupiter	Saturn	Uranus	Neptune	Pluto
Distance from Sun in cm	10	18	25	39	130	240	490	760	1000
Diameter in cm	1	2.4	2.6	1.4	30	25.2	10.4	10	0.8

Remember that the Sun is 110 times bigger than Earth. You will need to decide how to show this on your model.

1 Use the information in the table to draw and then cut out the planets to scale. Use information from this chapter to colour them or to add rings if you want to.

2 Find a suitable place to build your model. It could be a long passageway or a wall outside.

3 Stick or hang your planets so that they are more or less the correct distance from the Sun.

4 Label the parts of the Solar System on your model.

Unit 3 The Earth and the Moon

You have learned that some planets in the Solar System have moons. Moons are natural satellites that orbit around the planets they are 'attached' to. Our Moon is the closest space object to the Earth. We can see the Moon in the night sky because it reflects light from the Sun. You cannot usually see the Moon during the day, because it is light, but it is still there!

Astronomers are discovering new moons all the time so you will find different numbers of moons given for the planets in different books, depending on when they were published.

➜ **Figure 14.8**
The full Moon seen from Earth

Moon movements

The Moon revolves around the Earth in an elliptical orbit once every 29½ days. At the same time, the Moon turns around, or rotates, on its own axis. The rotation of the Moon keeps the same side facing Earth. This means that we only ever see one side of the Moon from Earth. The side that we never see is called the dark side of the Moon.

You can see the position of the Moon at different times in its orbit, in Figure 14.9.

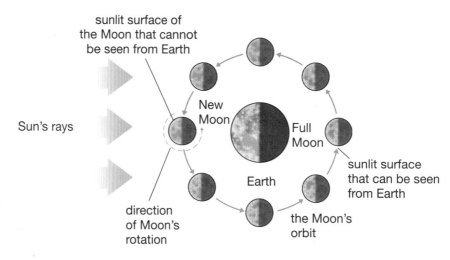

sunlit surface of the Moon that cannot be seen from Earth

Sun's rays

New Moon

Full Moon

Earth

sunlit surface that can be seen from Earth

direction of Moon's rotation

the Moon's orbit

➜ **Figure 14.9**
The shape of the part of the Moon we can see at different times

The phases of the Moon

If you look at the Moon on different nights, you will see that it seems to change shape. These changes are caused by the Moon's position as well as the positions of the Sun and the Earth. One half of the Moon is always in sunlight but we cannot always see the entire sunlit section because of the Moon's position. The only time we can see a whole sunlit face of the Moon is during Full Moon. At New Moon the sunlit face is facing away from Earth and we cannot see the Moon at all.

Figure 14.9 shows the eight main positions of the Moon during its orbit. The pictures in Figure 14.10 show you how the Moon appears in the sky at these eight points. We call these the **phases of the Moon**. The Moon goes through these phases every 29½ days. The first stage is called waxing because the Moon seems to be getting bigger from New Moon to Full Moon. The next stage is called waning because the Moon seems to get smaller from Full Moon to the next New Moon phase.

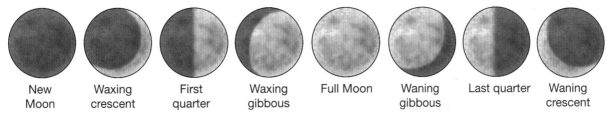

| New Moon | Waxing crescent | First quarter | Waxing gibbous | Full Moon | Waning gibbous | Last quarter | Waning crescent |

↑ **Figure 14.10** Each Moon phase has a special name.

Activity 14.5 Observing and recording Moon phases

1 Keep a record of Moon phases for about four weeks. Start tonight.
   a) Carefully draw 28 empty circles on a piece of paper, arranged in four rows of seven circles each, for example. These will represent what you see of the Moon.
   b) Look at the shape of the Moon in the night sky each night, and colour in yellow the shape that you can see, on one of the circles in your diagram.
   c) Write the date on each circle.

2 At the end of the month, try to spot and label when the eight main phases of the Moon occurred on your diagram.

3 Will the Full Moon occur on the same date every month? Give reasons for your answer.

4 The start of the Muslim holy month of Ramadan is closely linked to the phases of the Moon. Find out how.

Unit 4 Blocking the light

↑ **Figure 14.11** These people are getting ready to watch an eclipse of the Sun. They have travelled a long way to see it.

↑ **Figure 14.12** Even during an eclipse, you should never look directly at the Sun, especially through a telescope or binoculars. The concentrated heat and light can blind you.

An **eclipse** happens when an object in space moves in front of another object and blocks out the light from the Sun. On Earth we experience two types of eclipses: **solar eclipses** and **lunar eclipses**.

Solar eclipses

Sometimes the Moon moves directly between the Sun and the Earth. In this position, the Moon blocks out the light from the Sun. Parts of the Earth are then in the Moon's shadow and no sunlight can reach them.

Two or three times a year we can see solar eclipses from Earth. Because the Moon is much smaller than the Earth, it casts a shadow on a small area and the eclipse is seen in specific places for a short time only. When the Moon only blocks out part of the Sun it is called a partial eclipse. When the Moon completely blocks out the Sun, it is called a total eclipse and the area on Earth which is in the Moon's shadow turns dark in daytime.

In the past, people were afraid of eclipses. The Chinese believed that an eclipse was caused by a dragon trying to swallow the Sun. They would make as much noise as possible during an eclipse to scare the dragon away.

Lunar eclipses

Lunar means 'of the moon'. Sometimes the Earth is in a position between the Sun and the Moon and it blocks the light to the Moon. The Moon is in the Earth's shadow and it seems to disappear because no sunlight can be reflected off it.

When the Earth only blocks out some of the sunlight, only part of the Moon disappears from view. This is called a partial lunar eclipse. When the Earth totally blocks the sunlight from reaching the Moon, the whole Moon disappears from view and it is called a total lunar eclipse.

Activity 14.6 Labelling eclipse diagrams

1 Make outline copies of these two diagrams into your book.

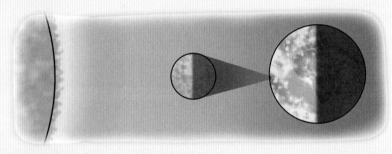

Sun
Moon
Earth
night
day
Moon's shadow
dark area on Earth

↑ **Figure 14.13** A solar eclipse

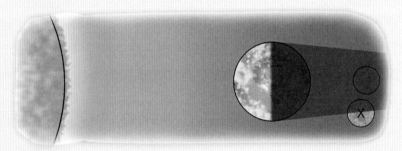

Sun
Moon
Earth
night
day
Earth's shadow
dark area on Moon

↑ **Figure 14.14** A lunar eclipse

2 Choose the correct labels from the box beside each diagram and label the diagrams to show what happens during a solar and a lunar eclipse.

3 If the Moon was at position X during a lunar eclipse, what type of eclipse would be experienced? Why?

4 Use a torch and two spheres of different sizes (a large one for the Earth and a much smaller one for the Moon) to demonstrate what happens during a solar and a lunar eclipse. Write up how you did this.

Unit 5 The atmosphere

Human beings cannot survive in space without special equipment – the temperatures are too hot or cold, there is not enough oxygen for us to breathe and there is the danger of being hit by flying pieces of rock (meteors). But we can survive on Earth.

We can survive on Earth because our planet is in the right position in the Solar System to support life. If it was much closer to the Sun it would be too hot, if it was further away it would be too cold. Also, the Earth is surrounded by a layer of gases called the **atmosphere**. You cannot usually see the atmosphere but you can feel it around you when the wind blows.

The atmosphere is a very important layer and we cannot live without it because:

● it is made up of gases that humans and other living things use to breathe
● it holds water vapour, an important part of the water cycle
● it filters out harmful rays from the Sun and makes sure that the temperature on Earth does not vary too much.

Activity 14.7   Finding out about the atmosphere

Study Figure 14.15 on page 171 and read the information given about the atmosphere before you answer these questions.

1 Name the important gases in the troposphere.

2 How thick is the atmosphere?

3 Where is most of the air in the atmosphere? Why?

4 What happens to the temperature as you go higher in the atmosphere?

5 Which layer filters out harmful ultraviolet rays from the Sun?

6 How does the atmosphere stop or slow down meteors which get close to the Earth?

7 In which layers does most of the world's weather happen?

8 What do you think would happen if you opened the window of an aeroplane high above the Earth where the air is very thin?

9 How do you think space travellers who leave the Earth's atmosphere in rockets manage to stay alive?

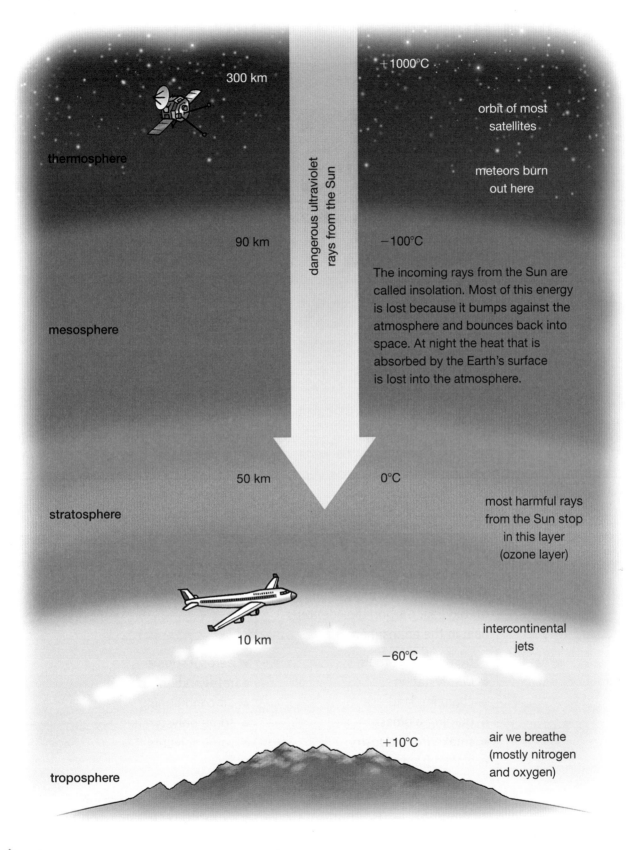

300 km

+1000°C

orbit of most
satellites

thermosphere

meteors burn
out here

90 km

−100°C

The incoming rays from the Sun are
called insolation. Most of this energy
is lost because it bumps against the
atmosphere and bounces back into
space. At night the heat that is
absorbed by the Earth's surface
is lost into the atmosphere.

mesosphere

dangerous ultraviolet
rays from the Sun

50 km

0°C

most harmful rays
from the Sun stop
in this layer
(ozone layer)

stratosphere

intercontinental
jets

10 km

−60°C

+10°C

air we breathe
(mostly nitrogen
and oxygen)

troposphere

↑ **Figure 14.15** The structure of the atmosphere

Unit 6 Humans and the atmosphere

Air pollution

➡ **Figure 14.16** We all need clean air to breathe and stay healthy. In some cities, the air is so dirty that people wear masks to try and stop the dirt from the air getting into their lungs.

Even though we cannot live without the atmosphere, some of the things we do can damage it because they produce poisonous and harmful gases. Some of these gases kill plants and damage trees. They may also get into clouds and become part of the water cycle where they poison water sources. Scientists believe that high levels of carbon dioxide in the atmosphere are causing the average temperature of the world to rise (global warming) and that this is affecting climate patterns. Chlorofluorocarbons (CFCs) are chemicals which are produced by some aerosols and used to keep fridges cold. CFCs in the atmosphere damage the protective ozone layer.

How do gases get into the atmosphere?

Carbon dioxide (CO$_2$)	Chlorofluorocarbons (CFCs)
This is the gas we breathe out. Trees and plants absorb this gas and give off oxygen. Carbon dioxide traps heat in the atmosphere.	This gas compound contains chlorine and destroys ozone.
<ul><li>burning fossil fuels</li><li>deforestation</li><li>changing land use</li><li>burning biomass</li><li>smoke from industry</li><li>smoke from car exhausts</li></ul>	<ul><li>aerosol sprays</li><li>refrigeration units</li><li>air conditioning</li><li>some polystyrenes</li><li>some solvents</li></ul>

Natural events can also add harmful gases to the atmosphere. When Mt Pinatubo in the Philippines erupted in 1991, it released 20 tonnes of sulphur dioxide into the troposphere. This caused the biggest climate change in the century – global temperatures dropped by 0.4 °C in 1992 as a result of the clouds of gas blocking sunlight.

Activity 14.8 Explaining global warming

1 Which gas is mostly blamed for global warming? How does this gas get into the atmosphere?

2 Work in small groups. Imagine you are scientists who have been asked to present a short radio programme on the effects of global warming. Use the information on the graph and maps below and what you learned in this chapter to prepare a short report. Read your report to the rest of the class.

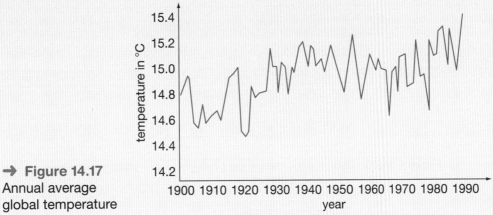

→ **Figure 14.17**
Annual average
global temperature

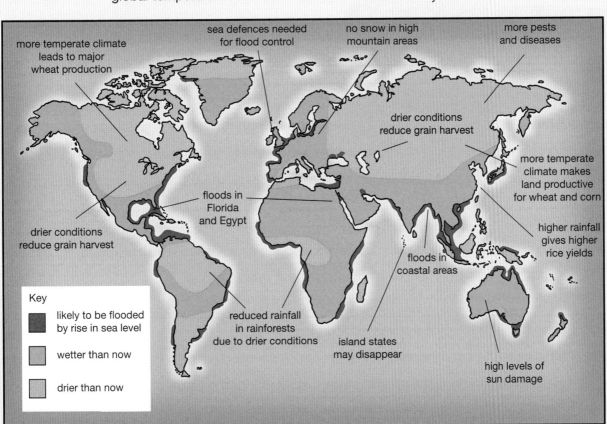

↑ **Figure 14.18** How global warming may affect the world

Chapter summary

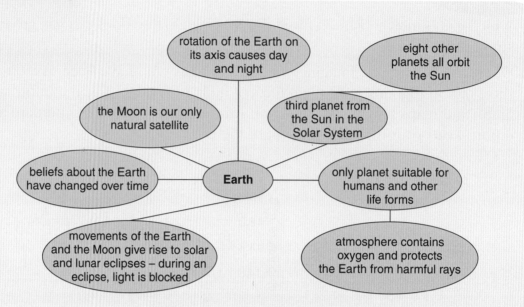

rotation of the Earth on its axis causes day and night

eight other planets all orbit the Sun

the Moon is our only natural satellite

third planet from the Sun in the Solar System

beliefs about the Earth have changed over time

Earth

only planet suitable for humans and other life forms

movements of the Earth and the Moon give rise to solar and lunar eclipses – during an eclipse, light is blocked

atmosphere contains oxygen and protects the Earth from harmful rays

↑ **Figure 14.19** This is a summary of what you have learned in this chapter.

Revision questions

1 Match each word in column A with the most suitable phrase in column B.

Column A	Column B
Sun	rocky planet third in order from the Sun
Earth	the star at the centre of our solar system which is the source of most energy on Earth
revolve	rising temperatures all over the world
eclipse	move in an orbit around the Sun
planet	partial or complete blocking of sunlight to Earth or Moon
ellipse	sphere of rock or gas that orbits the Sun
moon	oval shape
asteroids	natural satellite orbiting planet
atmosphere	rocks in space
global warming	the layer of gases that surrounds the Earth

2 Design an information leaflet about one of the planets of the Solar System (other than Earth). Your leaflet should include information about how this planet is different from Earth.

Glossary

A

absorbent   able to take in liquids easily

acid   a chemical that has a pH of less than 7

adaptation   a characteristic that makes something suitable to a particular environment

alkali   a base that will dissolve in water

aquatic   in water

astronomy   the study of the Sun, Moon, stars, planets and other objects in space

atmosphere   the layers of air that surround the Earth

B

base   a chemical that has a pH of more than 7

biome   an environment with certain weather conditions and plants

boiling point   the temperature at which a substance boils

brittle   hard but easily breakable

C

carnivore   an animal that eats the flesh (meat) of other animals

cell membrane   the thin layer around the contents of plant and animal cells

cell sap   the liquid found inside the vacuole of plant cells

cell wall   a thick cellulose layer found around the outside of plant cells

characteristic   a feature that is typical of a particular thing and that makes it different from other things

chemical change   a change which results in new substances being formed

chlorophyll   a chemical found in plants which gives them their green colour

chloroplasts   small structures in plant cells that contain chlorophyll

chromatography   the process of splitting colours into separate parts

classification   ordering of plants and animals into groups

community   a group of animals and plants that live in the same place

components   parts

compound   a chemical substance made of two or more elements joined together

compressed   squeezed or squashed by force

conclusion   something you decide based on observations or the information that you have

condensation   the change from a vapour or gas to a liquid as a result of cooling

conduction   the ways in which heat or electricity moves through solid substances

conductor   a material that allows heat or electricity to pass through it

consumer   animal which eats other animals or plants in an ecosystem

contact force   force exerted between two objects which are touching each other

continuous data   information that has fractional, or in-between, values

contraction   becoming smaller in size or shorter, usually as a result of cooling

convection   the way in which heat moves through a liquid or gas

corrosive   a substance that will cause a slow change on the surface by eating it away

culture   the beliefs and way of life accepted and shared by a group of people

cytoplasm   the liquid outside the nucleus of a cell in which the other cell structures are found

D

decanting   pouring from one container into another

decomposer   bacteria and fungi which break down other substances

dependent variable the factor in an experiment which changes depending on what you do

development the process by which plants and animals grow and become more complicated

diffusion spreading out through air or liquid

discrete data data which has only whole values

dissolving the process by which substances are taken up into liquids to form a solution

distillate the liquid condensed from vapour in distillation

distillation purifying a liquid by boiling it and then cooling the vapour so that it condenses back into a pure liquid

ductile able to be pressed or pulled into thin wire

E

eclipse a event in which the Moon or the Earth blocks out light from the Sun

ecosystem all the plants, animals and non-living things in an environment

effect the result of something

elastic returns to its original shape when the force on it is removed

electric current the flow of electricity through a circuit

electrical circuit a closed system, with a source of electricity, around which current can flow

electromagnetic waves a family of waves, including radio waves, light and X-rays, that can transfer energy through space

emulsion a mixture of liquids which do not dissolve in each other

environment the surroundings in which a plant or animal lives

evaporation when a liquid changes into a gas, normally as a result of heating

excretion getting rid of waste products

expansion spreading out or becoming bigger, normally as a result of heating

F

fact something that is true, or proven

fair test experiment or investigation in which all but one of the variables are kept the same to control the conditions

features another word for characteristics, the things that make something different from other things

field of view what you can see when you look through a microscope (or telescope or binoculars)

filter a very fine sieve used to separate a solid from a liquid

filtrate the clean substance that has been filtered

flexible able to bend or move

food chain a relationship between living things that depend on each other for food energy

force any kind of push, pull or twist

freezing point the temperature at which a substance freezes

fuel a material that is burned to produce heat or energy

function what something does, or what is it used for

G

gas a substance, such as air, which has mass, but no fixed shape

genus a group to which plants or animals with very similar characteristics belong

growth getting bigger, increasing in size

H

habitat the place where a plant or animal lives

herbivore animal which eats only plant materials

I

immiscible unable to mix or blend

in series connected in a straight line

independent variable the factor which you can change in an investigation or experiment

indicator   a substance used to measure changes (in pH value)

inherited   passed on from your parents

insoluble   unable to dissolve in a liquid

insulator   substance which reduces or stops the movement of heat or electricity through it

invertebrates   animals without a backbone

ion   an atom or group of atoms with an electric charge

K

key   a diagram or set of questions to identify a living thing

kingdoms   the five big groups into which all living things are classified

L

lifecycle   the series of changes that a living thing goes through from birth to death

liquid   a substance that can flow, like water

lunar eclipses   eclipses in which the Earth blocks the light from the Sun so that part of the Moon is hidden

M

magnify   make something look bigger

malleable   able to be shaped or bent

matter   the material that everything is made from

maturity   adulthood, being fully grown

measuring   finding out the amount of something; we can measure size, mass, volume, time, temperature and many other quantities

melting point   the temperature at which a substance melts

miscible   able to mix

mixture   combination of two or more substances that can easily be separated

model   a simplified version that is used to explain how the real thing behaves

movement   motion, changing position

N

natural materials   materials found naturally on Earth, not made by humans

neutral   a substance with a pH of 7, neither acidic nor basic

neutralisation   the process of mixing acids and bases to make a neutral substance

non-contact force   a force that acts without touching the object it is acting on

non-renewable   unable to be replaced in human lifetimes

nucleus   part of a cell, the control centre of the cell

nutrition   eating and digesting your food

O

observation   something you notice by looking closely

offspring   young, babies

omnivore   animal that eats both plants and other animals

opaque   unclear, not see-through

opinion   what someone thinks

orbit   the path that planets take around the Sun

organ   a collection of tissues that work together to carry out particular functions

organ system   a group of organs that work together to perform a function

organism   living thing

P

particle model   scientific idea that all matter is made up of particles arranged in different ways in solids, liquids and gases

pH scale   scale of measurement to show how acidic or basic a substance is

phases of the Moon   the changes that we see in the shape of the Moon through the month

photosynthesis   the process by which plants turn sunlight into food

physical change   a change that is easily reversed and in which the substances themselves do not change

population   the number of plants or animals of a particular species living in a habitat

pressure   a force spread over an area

producer   the plants that are at the start of every food chain

properties   the qualities or characteristics of a substance or organism

pure substance   substance that cannot be physically separated

R

radiation   the way in which heat moves through space, as waves

renewable   able to be replaced easily and quickly

reproduction   making new versions of themselves, producing offspring

residue   the remains of something, the pieces left behind

resistance   the ability to reduce the flow of electricity

respiration   using oxygen to get energy from food

S

saturated   full, unable to take in more

sensitivity   awareness of, and reaction to, the surroundings

sieve   a mesh filter

solar eclipses   eclipses in which the Moon blocks the light from the Sun so that all or part of the Sun is hidden from Earth

solid   any substance that keeps its shape

solubility   how well a substance dissolves in a liquid

soluble   able to dissolve in a liquid

solute   the substance which is dissolved in a solvent

solution   a liquid in which something has been dissolved

solvent   the liquid part of a solution

specialised   adapted to do particular jobs

species   one kind of animal or plant

states of matter   the different ways in which matter can appear – solid, liquid and gas are all states of matter

stiff   difficult to bend

sublimation   the process of changing from a gas to a solid without becoming a liquid

survey   finding out about something by asking questions or counting

suspension   solid materials floating in a liquid, not dissolved

synthetic materials   materials made by humans

T

terrestrial   on land

tissue   a group of similar cells that are organised to carry out the same function

transfer   move from one substance or place to another

translucent   allowing light to shine through, but not clear

transparent   see-through

U

unique   one of a kind, special or different

V

vacuole   space in a plant cell filled with cell sap

variations   small changes

vertebrae   the bones in the backbone (spine)

vertebrates   animals with a backbone

vibrate   shake or move from side to side

W

waterproof   unaffected by water

water-repellent   does not absorb water, but not completely waterproof

water-resistant   does not get wet, but not waterproof

work   the transfer of energy from one thing to another